Franziska Sörgel
Emotional Drivers of Innovation

Science Studies

Franziska Sörgel is a Post-Doc at the Institute for Technology Assessment and Systems Analysis (ITAS) at the Karlsruhe Institute of Technology. The cultural anthropologist received her doctorate at Humboldt-Universität zu Berlin in 2023. Her research focuses on anthropological perspectives of emotional evaluation and decision-making and alternative forms of impact assessment.

Franziska Sörgel
Emotional Drivers of Innovation
Exploring the Moral Economy of Prototypes

[transcript]

Berlin, Humboldt-Universität zu Berlin, Kultur-, Sozial- und Bildungswissenschaftliche Fakultät, Dissertation, 2023, 'The Moral Economy of Prototypes or How Emotions Drive Innovation – An Ethnographic Exploration in the Innovation-Making Field'.

Publication costs funded by the Institute for Technology Assessment and Systems Analysis (ITAS), Karlsruhe Institute of Technology (KIT).
Text editing supported by gender equality funds from Humboldt-Universität zu Berlin 2022.

Bibliographic information published by the Deutsche Nationalbibliothek
The Deutsche Nationalbibliothek lists this publication in the Deutsche Nationalbibliografie; detailed bibliographic data are available in the Internet at https://dnb.dnb.de/

This work is licensed under the Creative Commons Attribution-ShareAlike 4.0 (BY-SA) which means that the text may be remixed, build upon and be distributed, provided credit is given to the author and that copies or adaptations of the work are released under the same or similar license.
https://creativecommons.org/licenses/by-sa/4.0/
Creative Commons license terms for re-use do not apply to any content (such as graphs, figures, photos, excerpts, etc.) not original to the Open Access publication and further permission may be required from the rights holder. The obligation to research and clear permission lies solely with the party re-using the material.

First published in 2024 by transcript Verlag, Bielefeld
© Franziska Sörgel

Cover layout: Maria Arndt, Bielefeld
Cover illustration: Mathias Schmidt, Berlin
Proofread: Kelly GmbH
Printed by: Majuskel Medienproduktion GmbH, Wetzlar
https://doi.org/10.14361/9783839471470
Print-ISBN: 978-3-8376-7147-6
PDF-ISBN: 978-3-8394-7147-0
ISSN of series: 2703-1543
eISSN of series: 2703-1551

Printed on permanent acid-free text paper.

Contents

Acknowledgement 7

List of Abbreviations 9

List of Figures 11

I. The Sensitivity of The New 13

II. Methodological Approaches and Empirical Analysis 25

Theory – Thinking, Feeling, and Acting in the Moral Economy

III. From Problem to Possibility
    Imagination, Experience, and Emotion 49

VI. Innovation-Making
    The Construction of Value 81

Empiricism – A System of Emotional Forces Around Innovation

V. The Imaginative Remedy 121

VI. Premises and Other Problems 141

VII. Emotions as Valuta 167

VIII. The Moral Economy of Different Intentionalities 185

References 195

Acknowledgement

This book derives from my doctoral thesis, which I successfully defended at Humboldt-Universität zu Berlin in the summer of 2023. It represents the culmination of extensive contemplation on how ideas originate and evolve through emotions and experience, particularly within biomedical technologies. It posits that ideas and their later materialisation are an expression of what one feels and cares about (as a result of individual experience), be it oneself, the environment, or the future and that this is a relationship that takes place in mutual dependence, whether in the context of biomedical technologies or beyond, as I realised. The book's cover serves as an interpretation of this: it depicts multiple hands, symbolising the diverse individuals who contribute and influence the developmental process through their standards.

Moreover, this book is not only the result of interdisciplinary studies across various disciplines of the humanities and social sciences under the guidance of exceptional mentors but also of years of contemplation and exploration within the field of technology development. This book is situated within the realm of Science and Technology Studies and contributes to both. In the realm of Science Studies, this book examines the interplay of different disciplines and their inherent norms and beliefs in the act of fabrication. Like eminent scholars preceding me, I illustrate that researchers are not impartial agents but rather influenced by the established structures in which they operate. Simultaneously, this book contributes to Technology Studies by examining the genesis of ideas as prototypes within the context of their conception and the surrounding world and structures, such as incubators, into which these ideas are born.

Writing this book has been a transformative journey for me. I have gleaned invaluable insights from the texts I have read, the conversations I have engaged in, the places I have visited, the people I have observed, and the process of writing—before, during, and after a pandemic. I am fortunate to have worked and conducted research at the following universities during my doctorate: Technical University Munich, Humboldt-Universität zu Berlin, European University Institute in Florence, and Technical University Berlin. I am profoundly grateful to have studied at these remarkable universities alongside and under the guidance of generous individuals who dedicated their time to engage in meaningful conversations and exchange ideas.

I want to thank Jörg Niewöhner for helping me organise a study abroad experience at Goldsmiths College, University of London, during my master's programme. Studying there marked a new genuine enthusiasm for my endeavours and denoted the start of this journey. Therefore, I thank Martin Savransky for supporting me in the idea for my doctoral thesis and encouraging me, as well as Martin Reinhart and Michael Guggenheim for granting me a position as a doctoral candidate and dedicating their time to mentoring me. For helpful comments and inspiring conversations, I would also like to express my appreciation to Knut Blind, Tanja Bogusz, Lorraine Daston, Sascha Dickel, Benno Gammerl, Angela Graf, Monica Greco, Jana Heinz, Georg Jochum, Wolfgang Kaschuba, Sebastian Pfotenhauer, Arnout Van De Rijt, Marsha Rosengarten, Alexander Wentland, Alex Wilkie as well as to my colleagues from *Zentrum Emanzipatorische Technikfoschung* (ZET), and former classmates and friends Julia Bernighausen, Nora Erdmann, Judith Hartstein, Teresa Isigkeit, Mena Lüsse, Lew Mac, Kurt Rachlitz, Adrien De Sutter, who supported me in various ways throughout this undertaking.

My thanks go to transcript, especially Pia Werner, the editorial and design team, for supporting the publication of this book and to Mathias Schmidt, a Berlin artist, who kindly made his screen print available as the cover.

I extend my heartfelt appreciation to my family for their support during these years in manifold ways. My parents and brothers, with whom I had regular conversations, have consistently been a driving force for this work. My profound appreciation goes to Nikolai who has supported and encouraged me in all aspects of life since our encounter, which has ever since meant a revolution. Finally, I thank our son Valentin whose mere presence acted as a catalyst for completing this work and once again has confirmed that imagination is inherently human.

I would like to express my gratitude to the Institute for Technology Assessment and Systems Analysis (ITAS) at Karlsruhe Institute of Technology (KIT) for their generous financial support in publishing this book.

List of Abbreviations

AI	Artificial Intelligence
BMBF	Bundesministerium für Bildung und Forschung (German Federal Ministry for Education and Research)
BMWi	Bundesministerium für Energie und Wirtschaft (formerly German Federal Ministry for Economic Affairs and Energy)
CE	Conformité Européenne (security seal in Europe)
CEO	Chief Executive Officer
CMS	Centers for Medicare & Medicaid Services (in the US)
CTA	Constructive Technology Assessment
DAI	Distributed Artificial Intelligence
DIY	Do It Yourself
EC	European Commission
EU	European Union
FDA	Food and Drug Association (in the US)
GT	Grounded Theory
ICU	Intensive Care Unit
IoT	Internet of Things
IP	Intellectual Property
LISREL	Linear Structural Relations
MIT	Massachusetts Institute of Technology
MOX	Museum of Failure
OEM	Original Equipment Manufacturer
POCD	Postoperative Cognitive Dysfunction
R&D	Research and Development
RRI	Responsible Research and Innovation
STS	Science and Technology Studies
TED	Technology, Entertainment, Design
VR	Virtual Reality

List of Figures

Figure 1:	Experience-Imagination Interrelations	57
Figure 2:	From Experience to Evaluation	68
Figure 3:	The Emotive Corporate Culture	91
Figure 4:	Iteration Loops of Expectations During the Development Process	95
Figure 5:	The (Linear) Narration Pattern	99
Figure 6:	Evaluation Interactions Between the Prototype and the Individual	108
Figure 7:	The Non-Linear Evaluation Interactions	113
Figure 8:	Inside the M.lab – Container One aka Utopia	123
Figure 9:	Sketch by Hendrik explaining the Prototype's Genesis	128
Figure 10:	Third Prototype of the Insole	129
Figure 11:	Trying Out the Insole in an Orthopaedic Shoe	131
Figure 12:	Exploration Phase One at the Creative Space	137
Figure 13:	Hydrocephalus Valves in Different Stages of Development	140
Figure 14:	Explaining the Idea – Finding a Lingua Franca	143
Figure 15:	Jan's Perspective: From the Idea to Selling the Product	173
Figure 16:	In Preparation for the Demo Day during the COVID-19 Pandemic	176

I. The Sensitivity of The New

In the ethnographical examination of entrepreneurial-creative places such as makerspaces, incubators, or living labs, I frequently encountered assertions and explanatory patterns imbued with emotion. Interestingly, these narratives often seem to evade a conscious acknowledgement of their emotional nature. Instead, participants describe experiences as *magical moments, visions of a brighter future, a flourishing corporate culture,* or *serendipitous discoveries.* When explicitly asked about the emotionality of the innovation process during my research, respondents displayed hesitation, sometimes coupled with expressions of wonder, and asked what I meant or what emotions I was talking about. Perhaps the most frequently posed question in the context of my research was: 'What kind of emotions do you mean?' to which I was expected to pinpoint specific emotions like 'fear', 'happiness', or 'sadness' to elucidate my research question. At this point, wonder usually arose on both sides. I intended to investigate emotions without constraining them within predetermined categories. However, some participants found this approach too abstract, prompting a desire for a more concrete framework. Consequently, I promptly and significantly adjusted my approach.

This anecdote holds particular relevance to the broader research context, especially when investigating the role of emotions as an 'impact factor'. To unravel the influence of emotions on innovation processes or the shaping of prototypes, delving into the (sometimes) imperceptible in the materialisation process becomes imperative. Providing a predefined frame of reference would not align with the research question, as it might inadvertently impose limitations. These limitations could hinder participants from discussing what may be retrospectively recognised as emotional aspects, steering conversations towards what fits within the predetermined frame.

Furthermore, this discovery underscores the intricacies inherent in exploring emotions, shedding light on a broader challenge. Despite the potential for collective character, emotions are frequently subjectively experienced, rendering their scientific assessment complex. Anything rooted in subjective experience or inherently resistant to objectivity has historically faced discrediting and dismissal as non-scientific. The objectification of science, entrenched in the process of rationalisation with a longstanding tradition(e.g. Daston & Galison, 2007; De Sousa, 1987), further complicates the evaluation of emotions. Consequently, even within an innovation process, inherently a knowledge generation endeavour, participants in innovation-making may not openly accord significant importance to emotionality, particularly when adhering to presumed linear developmental trajectories.

Hence, scrutinising emotions in a knowledge production process like innovation often encounters scepticism, lacking serious consideration. This scepticism is particularly pronounced in the realm of innovation research, typically approached quantitatively by economists or through models rather than qualitatively. Despite this, there is a noticeable paradox: market participants frequently and consciously infuse the concept of innovation with emotion in its external portrayal. This disjuncture is striking. On the one hand, those engaged in knowledge generation may not consciously acknowledge emotions or only do so after persistent efforts. On the other hand, they deliberately leverage and manipulate emotions in the marketing of nascent products. From these ambivalent observations, my research question takes shape: *To what extent do the emotions of participating actors matter in the materialisation of an idea, and how can these emotions constitute a prototype's 'moral economy'*[1]*, providing insights into the society in which it emerges?*

Hence, this book investigates the manifold modes of communication integral to the evolution of a prototype, commencing with its initial conception. Central to this exploration is an emphasis on the emotions experienced by the actors throughout the development process. The germination of an idea for a potential innovation typically springs from a problem embedded in the everyday lives of individuals, grounded in specific experiences. In my inquiry, I meticulously trace the genesis of the idea, contending that it is inherently and already emotionally constituted. The theoretical framework posits that the interplay between *experience* and *imagination*, manifested as reflective forms of interaction, gives rise to *emotion*. Yet, establishing a definitive sequence, whether experience precedes imagination or vice versa, proves elusive. The act of consciously perceiving, observing, or memorising an experience in daily life generates emotional resonance in inventors, empowering them to envisage potential solutions. Consequently, a triad emerges—comprising *experience*, *imagination*, and *emotion*—intertwined and mutually dependent, though not bound by a linear temporal sequence.

Through the involvement of additional actors and team members within the incubator or makerspace, the trajectory from the conception of an idea to the circulation of a prototype becomes a collaborative process. These individuals often carry distinct expectations shaped by their experiences, thereby introducing an emotional disposition that undergoes processing and negotiation within the iterative loops of prototype development. The significance of these expectations lies

---

1   A site or forum of a prototype's negotiation process (see Chapter III).

in their capacity to offer enduring insights into evaluation patterns and categories within both innovation and scientific research. Moreover, adopting an alternative approach, marked by the emotionalisation of the innovation process, serves as a gateway to alternate narratives concerning motivations for innovation. Contrary to the frequently propagated narrative of linear developmental paths, this study highlights that ruptures emerge and become conspicuous during iteration loops, constituting a substantial aspect of innovation development. These ruptures manifest initially through the erosion of idealisation at the conceptualisation of the idea and continue to evolve in diverse ways.

This research scrutinises emotions and team dynamics through qualitative ethnographic methods, encompassing participatory observation, interviews, 'work-alongs' (in-situ research linked to interviews), and comparative analysis.

To begin with, Chapter II discusses the methodological approach and how the empirical data was collected and analysed, as individual interview quotes already contribute to the theory section in relevant passages. This approach links the theory with empiricism from the beginning, creating a nexus among the chapters that ensures it is clear who participated in the survey and whose words are being quoted.

Accordingly, the first subchapter discusses how I entered the innovation-making field. As with any empirical study, there is an inherent risk of bias or one-sidedness. In addition to avoiding one-sidedness in the present work, my main concern was maintaining flexibility concerning the numerous perspectives in the field of innovation and comprehending the many specialised languages spoken by diverse actors. The empirical work required me to remain in the interplay between myself and the research subjects and adopt a reflexive attitude. Accordingly, I combined grounded theory (GT) with John Law's approach of 'method assemblages' (Law, 2004: 13), aiming to avoid a strictly classical methodological approach. This approach fosters openness in the researcher, enabling the capture of diverse realities without prioritising specific perspectives. This can create the impression of an external 'messiness' (Law, 2004: 18), which, however, turns out to be a complex mosaic for presenting a holistic picture of multifaceted fields. At the latest, the reflexive approach proves its worth when tracing the changing narratives of the same person or different narratives about an artefact simultaneously. These moments are tense, perplexing, and often loaded with emotions, which is precisely why these passages are interesting for research. In subchapter 2.3, I present the empirical sites where I conducted my research. The first site is an incubator where I could get to know two projects, namely *Feety* and *Ellie*, both of which involved developing physical prototypes.

The second site is a makerspace, where I looked at general structures, while the third is a creative space, which takes a different approach to evolving an idea, although the names resemble each other. The fourth site was an established company that has dominated an entire world market with one novelty. To round this off, I spoke with an innovator who has also been a private investor and thus fulfilled a notable dual role. All sites and interlocutors contribute to observing and understanding the origins of ideas, most of which are related to medical technology developments. Although the theoretical assumption goes beyond the reference to medicine or medical technology, due to the care aspects inherent in the practice of medicine, emotional factors are also involved to a certain extent, which is why it initially seemed more accessible to create an approach to emotions in this area. This assumption is helpful for initial theses but not necessary to trace the origins and general development of ideas.

I ethnographically researched the various interlocuters' work and ideas using different methods. I conducted interviews, observed them at work whenever I could, photographed their activities and emerging prototypes and had them draw work processes.

Subchapter 2.5 refers to the problems encountered before, during, and after the survey in the sense of a reflexive attitude towards the field and the researcher. Particularly noteworthy at this point are the issues of confidentiality, which especially applies to the area of innovation, and the global crisis surrounding the COVID-19 pandemic, which considerably impacted my research. The hurdles of confidentiality and concerns about innovative ideas being exposed by an external researcher, coupled with the challenges of a pandemic that forced teams of developers into isolation and home offices or closed entire creative venues, presented me with particular challenges and significantly limited the possibilities for data collection.

Chapter III serves as the inaugural theoretical segment of this book, meticulously examining an idea's genesis from its phenomenological roots. The central inquiry revolves around the interplay of imagination, experience, and emotion, culminating in what is later termed a 'moral economy'. This concept functions as a dynamic site of negotiation, proving pivotal in subsequent processes of collectivisation.

In retracing the origins of an idea, as expounded in subchapter 3.1, imagination emerges as a critical factor for shaping the future, a space where possibilities of change unfold. Hypotheses and 'what if'-scenarios become vehicles for exploring this imagined terrain. Franz Brentano calls this ability *mental force*. Drawing on his pre-phenomenological insights, the work elucidates how imagination, grounded in personal experiences, establishes a reciprocal relationship with the world. This interplay involves a continuous dynamic wherein the present self mediates between the past self and the external world. Brentano terms this ability to relate to something both inside and outside as *intentionality [Ger. Intentionalität]*. The physical existence of an object becomes secondary to its mental existence; an object becomes real through imaginative envisioning. This unleashes a creative force essential for idea development, with experience playing a supportive role in releasing imaginative power.

Subsequently, subchapter 3.2 addresses the origin of ideas and the nexus between experience and creativity, drawing on the pragmatist approaches of William James and John Dewey. These approaches integrate feeling into the broader context of thought and action, overcoming the subject-object dualism of modernity through radicalempiricism. By creating a holistic picture of the body and the environment, Dewey, in particular, underscores the relevance of the sensed or subjective aspects. Conscious experience, according to Dewey, materialises through active interaction with the world, fuelling conscious doing and creativity. Everyday actions take precedence in the perspectives of these pragmatists, serving as the fertile ground from which ideas originate.

Moving forward, subchapter 3.3 delves into the emergence of emotions through interaction with a created artefact and their subsequent transmission. Emotions, integral to our modes of communication, exert influence over the process of knowledge production within the communicative realm—an influence that transcends individual occurrences and extends into collective dimensions. Grounded in the foundational premise of communication as fundamentally interpersonal, this subchapter elucidates the incorporation of artefacts into the intricate dynamics of communication. Here, these artefacts assume the role of a projection screen, facilitating the expression and evaluation of emotions between interacting subjects. Within the context of the historical trajectory of emotions in scientific discourse, a distinctive emergence is discerned—the 'scientific self' (Daston & Galison, 2007, p. 191 ff.) assumes agency within this communicative network, actively contributing to the ongoing generation of knowledge through the mediation of feelings, assumptions, and experiences. Emotions, historically relegated to scientific pursuits, underwent a transformative shift with the advent of the *Linguistic Turn* in the early 20th century. The pragmatists of the 20th century played a pivotal role in elevating the status of emotions within the scientific domain. Disciplines like psychology engaged in a rigorous scientific exploration of emotions, giving rise to novel perspectives that displaced the Cartesian view. Consequently, emotions shed the pejorative label of 'irrationality' and became integral subjects of contemporary sociological research. This paradigm shift paves the way for exploring a discernible 'grammar' inherent in emotions, a trajectory applied in my research to unravel the emergence of emotionality within the realms of ideation and innovation-making. Consequently, the crafted artefact transcends its status as a mere mental construct, evolving into an individual frame of reference for the inventor and an active agent within the multifaceted network of interacting actors. It is no longer confined to the realm of mental imagery; it exists as a tangible entity in the world, subject to negotiation and engagement.

In digging into the culmination of subchapter 3.4, we discern the artefact's transformative journey beyond a mere conveyor of emotions to becoming a vessel for values. Within these intricate interaction processes, team members engage in nuanced negotiations amongst themselves, exchanging not only knowledge but also their distinct moral perspectives. Those currently engaged with the artefact become integral actors in an economy that morphs into a vibrant arena, hosting their ideas, emotions, and judgements. This phenomenon, akin to what Lorraine Daston terms a 'moral economy' (Daston, 1995), profoundly influences the scientific landscape, shaping ruminations on what to think, the preferred subjects to explore, decision-making protocols, and the objects under scrutiny. These moral economies illuminate the intricacies of scientists' choices—why certain objects are selected, which explanations gain trust, and the habits or methods that are embraced or cultivated. Such insights serve as a compass, elucidating the emotional impact on key actors and offering glimpses into what they perceive as relevant. In the realm of innovation, diverse actors, despite their disparate origins and technical vernaculars, converge in a collective conversation, forging unity within their pluriverse.

The exploration of innovation structures unfolds in the second theoretical segment of this book (Chapter IV), building upon the preceding examination of experience, imagination, and emotion. Conceptualising innovation as a collective process, particularly in its practical manifestation, necessitates an investigation into its designated locales where the invention is practised in alignment with public understanding (subchapter 4.1). It is within these spaces that innovation, corporate culture, and creative cultures converge, giving rise to expectations. Initially, individual expectations surface, characterised by their ideal-typical nature, before being disseminated and shared within various creative realms for further refinement. The idealised idea encounters supplementary visions, providing a projection surface for diverse thoughts and desires that seek eventual realisation. As these expansions occur, the treatment of the prototype and its narrative undergoes a transformation. Subsequently, subchapter 4.2 examines the origin of an idea and its subsequent evolution, seamlessly weaving it into the object's narrative. Throughout this journey, the artefact remains a projection surface, accommodating the wishes, expectations, and future visions of an increasingly broad audience. These narratives transcend mere storytelling; they are performed and practised, imbued with emotional dimensions. Typically centred around a problem from everyday life, the narrative introduces an emotional component, transforming the problem into a call to action, seeking a solution. Within this context, invoked creativity emerges through the presented problem and the ensuing call to action. Notably, the narrative consistently operates as a founding myth, profoundly entrenched in emotional undertones. These evolving myths are replete with symbols, transcending mere textual content. The act of conveying a message employs a specific language and follows established rules, individually decoded by recipients, inevitably eliciting emotions. Consequently, the innovative idea or development thrives solely through its narrative, provided it aligns with the zeitgeist of society.

Ultimately (subchapter 4.3), values undergo a process of comparison, adaptation, and expansion within the context of collectivisation. Framed within the concept of a *moral economy*, a collective entity forms around the artefact, transcending its role as a mere transmitter of ideas and narratives. This collective not only conveys values but also actively shapes a shared morality, encompassing language, customs, and more. As previously suggested, these desires and ideas are inherently emotional, embodying a moral belief that guides how something should manifest its purpose or value at its core. The evaluation of an artefact's value often occurs in the context of success and failure. However, the categorisation of success and failure proves elusive and unpredictable, leading to the frequent invocation of the explanation of serendipity. Despite the potential influence of luck and chance being less significant than commonly assumed, the utilisation of these concepts reveals important insights. Contrary to popular belief, experimental spaces within creative environments such as incubators are tightly controlled, minimising the role of chance. Throughout these investigations, apparent contradictions abound. Whether attributed to chance or not, the field of innovation remains a mysterious black box, resisting easy insights for external observers. Innovation sites, despite being perceived as fragile and vulnerable, are often restrictive and opaque, guarded as precious entities by the entrepreneurial forces that govern them. This intentional closure adds layers of complexity and challenge, turning innovation into an enigmatic realm that demands closer examination. The intricate interpersonal dynamics within these innovation spaces, with their numerous gradients, offer compelling reasons to research their complexities.

Thus, the empirical exploration (Chapters V, VI and VII) substantiates the hypotheses posited in the theoretical framework through illustrative instances derived from the dataset. In this empirical study, the focus is directed towards the data material, drawing upon examples to validate the theoretical constructs. The investigations unfold within diverse settings, encompassing an *incubator* dedicated to biomedical technologies, a *makerspace*, and a *creative space*. Additionally, I observed interactions within an established *company* and gathered insights from a private entrepreneur who wears the dual hats of investor and innovator. The specific context revolves around three developments in medical technology, supplemented by a broader reference to advancements within the biomedical domain.

Chapter V examines the tangible unfolding of ideas—their inception and evolution—wherein the idealisation of these ideas emerges in response to less-than-ideal circumstances, fostering a (moral) purpose and motivation (subchapter 5.1). Ideas and dreams, construed as forms of imagination, aspire to manifest something grander and superior. Within a prototyping lab visited during my research, potential scenarios transform into utopian visions for inventors, serving as realms for self-realisation. The orientation toward the future in wishful thinking embodies an idealisation immune to disappointment due to its not-yet-finished nature. The utopia unfolding in the lab carries dual meanings: the prototype transforms into an imaginative utopia for inventors' wishful thinking, and the lab itself comes alive as a space for realising possibilities, thus cultivating a creative ambience in tune with the zeitgeist.

In subchapter 5.2, I reference the pragmatist triad of *thinking*, *feeling*, and *acting*, emphasising problem-oriented creativity that necessitates a conscious perception of a problem before it can be solved. Here, conscious perception is akin to *thinking*, and problem-solving aligns with *acting* in the triad. Both contemplating a problem and discovering a solution are emotionally charged aspects that subsequently evoke *feelings*, completing the triad. Throughout this chapter, individuals from a medical innovation incubator predominantly share their everyday experiences in the clinic. The meaningful discovery of problems reflects their exploration of the inventor's environment and the attention dedicated to it.

In the context of subchapter 5.3, along with the interviewees, I explore the significance of various emotions that foster intrinsic motivation in the innovation process, illustrated by the personal concern felt by the innovators based on their experiences. The innovators unanimously express that their work is fundamentally about making a difference through their inventions, positively influencing their immediate surroundings. The omnipresence of the phenomenon of world-changing action is evident in the available data. Subchapter 5.4 digs deeper into the aspect of inventors striving for *moral impact*. Building on the preceding subchapters, interview partners articulate how they navigate themselves and their environment, invoking concepts related to care. This extends to the world, which is dedicated to one's activity in correspondence with the environment, whether it involves medical technology or not. The guarantors share convictions that shape their purpose and beliefs, sustaining motivation over an extended period. Both elements contribute to the narrative presented when persuading others that the invention is worthwhile, urging them to invest in it, try it, or buy it.

In Chapter VI, the exploration focuses on the substantial challenges emerging within the process of idea generation, imbued with an emotional intensity that intricately shapes the decision-making processes. This encompasses consensus-building, the cultivation of shared values within a team, and the continuous emphasis on the central role of trust. The chapter explores the initial processes of reduction, where an idea, initially envisioned as perfect by the inventor, undergoes negotiations within the everyday dynamics of the surrounding collective.

As elucidated in subchapter 6.1, the origin of many problems often stems from the absence of a common language within a team, necessitating the development of an operational language through collaborative efforts. This proves challenging, particularly given the diverse origins and disciplines of team members, as observed in interviews and interactions. Experiences and resulting knowledge must be shared, and common goals must be developed to establish shared values. Practising one's speaking competence emerges as a crucial lesson in this context.

In subchapter 6.2, the focus shifts to another facet of collectivisation, where actors with diverse perspectives strive to overcome obstacles and unite into a cohesive entity. While a lingua franca can facilitate successful communication, it also underscores the intricate and multi-layered nature of the course of unification. The data reveals that problems are unpredictable and result from the plurality inherent in this collectivisation process, offering insights into the daily dynamics of innovation spaces.

Subchapter 6.3 maintains its focus on the processes of collectivisation, emphasising that overcoming conflicts is contingent on the existence of trust among team members. The fragility of these 'finding processes' in incubators becomes apparent, with the sustenance of trust being vital for relationships to endure or improve. Various perspectives on trust are explored, underscoring the delicate nature of team dynamics in incubators.

The final subchapter (6.4) delves into the potential danger of failure and the coping mechanism encapsulated in the saying, 'Fake it till you make it'. Interviewees share their strategies to persuade others, even when their ideas seem implausible. The saying becomes a deeply ingrained survival strategy in their daily lives, representing a marathon of innovation launch, where maintaining a composed demeanour and envisioning potential success as already achieved become essential until it materialises.

The final chapter (VII) unveils how the once ideal-based innovation process undergoes a sudden reduction for the purpose of out-licensing. The interviews not only bear witness to narrative adaptations but also reveal the emotionalisation of these narratives, strategically enhancing their market appeal. In subchapter 7.1, concrete examples of potential conflicts that could have been avoided are presented, notably focusing on intellectual property (IP) and the intricate questions surrounding legal ownership, interpretation, and evaluation of ideas. These conflicts underscore the evolving evaluation dynamics within the moral economy, where ideals gradually wane over the course of development.

Subchapter 7.2 captures the challenges faced by some interlocutors in asserting themselves against the expectations of financiers. Once the purpose of an innovation is formulated and a need is addressed, the crucial step involves marketing and emphasising its uniqueness. Here, emotions transform into commodities, creating a marketplace that reveals a distinct culture with clear hierarchies.

In the penultimate subchapter (7.3), the narrative shifts to how emotions are no longer exchanged within the confines of a team, incubator, or financiers but are 'performed' to construct a story on demo day. This narrative revolves around the idea and creation of a (not-yet-finished) artefact seeking a market in the audience. The concluding subchapter (7.4) accentuates the focus on reduction processes, inquiring about ruptures that disrupt the linear trajectory of innovation development but unfold on an emotional level. This marks a growing de-idealisation of development, where a rationalisation of feelings occurs to render a product marketable.

In this book, I extend the groundwork laid by social science and humanities researchers, particularly within Science and Technology Studies (STS). The prototype, as a research subject, has garnered increasing attention in the social sciences in the last decade (e.g. Dickel, 2019; Guggenheim, 2010; Guggenheim, 2014; Kelty et al., 2010; Nold, 2018). The objective of this research was to consider the prototype as an epistemic object: equally, as a materialised promise of the future or, through it, to recognise a form of experimentation inherent in the not-yet-finished thing. The same applies to the concepts of creativity and innovation, which have received no less attention. In a 'Do-It-Yourself (DIY) society' that is not only dedicated to repairing but to creating and has generously begun to share its knowledge and experience on the internet, the concept of creativity has now gained a great deal of importance in the literature (e.g. Florida, 2004; Moultrie et al., 2007; Reckwitz, 2017). In this context, creativity is increasingly perceived as a sensual-affective activity intertwined with innovating, serving as a prerequisite for the emergence of novel creations.

However, determining what qualifies as 'new' is not within the sole purview of innovators or inventors; it involves a negotiation process between society and innovators. The term 'innovation' has become pervasive, used almost carelessly by both scientists and entrepreneurs, possibly influenced by political calls emphasising innovation as a solution to grand challenges (European European Commission, 2010; Pfotenhauer, 2017). From Joseph Schumpeter's initial definition in 1911 to the present, innovation, described as 'destructive creation' (Schumpeter, 1942; Schumpeter et al., 2006), has repeatedly taken centre stage. Scholars from various disciplines have explored innovation, examining its social dimensions, success factors, and impact on various sectors (e.g. Briken, 2006; Curnow & Moring, 1968; Godin, 2017; John, 2012; Moultrie et al., 2007). Even entrepreneurs highlight corporate culture and collaborative joy as crucial success factors (Løw, 2018).

STS typically focuses on the sociality of technological artefacts, critically examining and predicting social developments. However, the analysis often revolves around Foucauldian power asymmetries within socio-material structures (e.g. Maasen, 2019), offering a view of structures but neglecting experiential content. The study of society in various fields, including science studies, the history of science, and sociology, is increasingly emphasising emotions, feelings, and affects as objects of study (De Sousa, 1987; Döveling et al., 2010; Hochschild, 2012; Illouz, 2017; Krüger & Reinhart, 2016). Yet, these elements have not been thoroughly explored as factors in technology development.

Research in the humanities and social sciences on the communicative function of prototypes is still nascent. Therefore, my book ventures into mostly unexplored academic terrain, utilising qualitative methods to approach the intersection of innovation and emotion. I scrutinise narratives of innovation through the lens of circulating prototypes, considering their occasional 'openness' in iteration loops as conducive to studying innovation processes. This contribution challenges the critical rationalisation of knowledge production processes within innovation and science research. By probing emotionalisation, my work questions evaluation criteria and patterns, providing a different perspective on knowledge accumulation processes in innovation research. As an STS researcher, I posit this book within Responsible Innovation and Technology Assessment, revealing the social mechanisms conditioning the connectivity and acceptance of technical development paths. This work illuminates how ruptures in innovation-making transpire, and prototypes ultimately align with a specific market logic, contributing to the re-enactment of innovation rather than preserving initially present social ideals, which are progressively neglected in the course of development. Through this alternative exploration via emotions, the work identifies moments that emphasise social ideals and highlights instances where they are disregarded.

II. Methodological Approaches and Empirical Analysis

I would like to begin by discussing the methodological approach of the ethnographic study. Although the respective cases with their material data are not discussed until Chapter V, individual interview excerpts are already included in the theoretical framing of this work, which represents an attempt to create a nexus between theory and empiricism and to link the material with the literature. Therefore, I describe the study's methodology, the methodological approach, and the handling of the material upfront to indicate how and under what circumstances the data were collected and the results obtained before commencing with the ethnography. Since most surveys took place during the COVID-19 pandemic in 2020 and 2021, this aspect cannot remain unnoticed as it was a significant obstacle when conducting the research.

To begin with (subchapter 2.1), I explain why I decided for the prototype to be the artefact of the investigation as it is more than a mere concept on the way to fulfilling a product. Further, I will elaborate on why the aspects of care in the context of medical prototypes initially seemed to be appropriate for approaching the field, although, as it subsequently emerged, the care aspect might generally be inherent in developing an idea once a problem has been discovered.

The next section (subchapter 2.2) explains the methodology of this study. Hence, I introduce grounded theory (GT) supported by a few arguments from 'Method Assemblages' by John Law (Law, 2004). Especially concerning marginalised data that threaten to get lost in large amounts of material, e.g. because they appear very subjective, this extension of GT seemed reasonable.

In subchapter 2.3, I present the research sites and introduce the informants in the field. I mainly spoke to people in the medical area from different makerspaces and incubators and to one private investor. I chose my informants based on their work with medical technologies or their strong emotional connection to their work in technology development.

To investigate the influences and effects of the actors' emotions that surround and influence the development paths of their prototype, e.g. in moments of decision-making, I used four different methods, in particular, to obtain various data materials following an ethnographic approach (subchapter 2.4), which includes (1) participatory observation, (2) in-situ narratives such as work-alongs, and (3) semi-structured interviews (both on-site and online) with innovators, incubators, and other actors in the workplace. Finally, (4) I analysed websites, videos, and brochures of the available prototypes and products.

The chapter closes with subchapter 2.5 and my reflections during and after the fieldwork, which is meant to shift the focus to the obstacles and general challenges in innovation-making.

2.1 Medical Prototypes as a Resource of Care

Prototypes as quasi-objects (Dickel, 2019; Latour, 1993) are particularly suited for the study because they are still in development and, therefore, not a finalised manifestation. Through their 'openness' and their resulting intermediate stage, they offer the possibility to compare their previous development with an original concept and, at the same time, examine the reasons why they are moving in the given direction of development. In this respect, the work contributes to innovation and science research against the background of the critical rationalisation of knowledge production processes whereby evaluation criteria and patterns are examined through the aspect of emotionalisation.

The research focused on medical technologies because it appears that the medical field, in particular, is often highly emotional, and moral reasons are frequently highlighted in connection with *care* when describing the original motives and reasons for the development of various medical aids/devices. The explanations around these artefacts seemingly include a posture that wants to 'fix' or 'heal' a given mental or physical deficit (Puig de la Bellacasa, 2017: 69; Tronto, 1993: 103). Thus, a simplified approach to emotions seemed feasible at first, as physical deficits are especially emotionally indicted because they are associated with pain, fear, shame, and the loss of dignity (Fineman, 1993: 19), to name but a few and, therefore, need special attention. However, aspects of care allow for a stronger focus on the sphere of emotions.

While emotions take on a greater role in the context of care, this connection does not only apply to technologies in the medical field since emotions are generally influential in technology development, as this book will highlight. On the contrary, regardless of the aspect of care, it can be stated that emotions require consideration as part of human cooperation. Although there are well-developed emotional motives in the field of medical technology development, it is challenging to approach these. As indicated in the course of this work, this is not due to emotions but rather to the field of innovation in general. Spheres that are labelled with the term 'innovation' are untransparent and challenging to enter, and it seems as if there are invisible walls of silence built around this theme. This difficulty is constantly apparent when describing the empirical study, whether concerning the approach, (not) surmounting hurdles to gain interviewees' favour, or how interviews are anonymised.

Despite the outlined challenges, the study still provides significant insights into innovation-making and the med-tech sector, primarily because qualitative methods are used so that the supposed 'marginal' can come to light. By 'marginal', I mean that access via emotions, whether narrated or observed, allows us to encounter subjective narrative forms about the motivations for innovating that do not otherwise occur, as the field is usually subject to rationalisation. In addition, emotions sustainably report on patterns of evaluation and categories that tell us something about the social issues of society.

In the following section, I present the applied methodology, which, while primarily representing the programme and approach of ethnography, is also exemplary of the inherent processes of the research field.

2.2 Methodology

To illuminate a diverse and sometimes vague research field such as innovation-making with questions that examine the emotional co-constitution in innovation-making, it is necessary for the researcher to frequently change perspectives and resort to various methods. It is precisely the subjectivity of emotions, or what the interlocutors describe as 'feelings', that opens up a vast arena of what can be researched in this field as an influencing factor in technology development. Additionally, I have found that feelings and emotions are often hidden in subordinate clauses and incidental situations that need to be crystallised. Therefore, implementing the approach of diversity in terms of methods and perspectives is necessary, not least to avoid potential one-sidedness or oversimplification. Further, the complexity of the data and its origins must be sorted and systematically arranged in the mosaic of its multiplicity, which GT helps to achieve.

GT is a proven methodology that was developed in the 1960s by the American sociologists Anselm Strauss and Barney Glaser which uses various qualitative methods to describe, classify, and interpret existing knowledge and, in this case, existing emotional states and treat them as social phenomena in social worlds including various actors. Doing so allows us to derive a theory from the material or to expand or modify already existing theories (Morse et al., 2021: 3–4). Generally speaking, GT as a methodological approach is helpful for explorative studies that are open to creatively spinning together the results into one bigger picture. Although it is primarily used to study interactions or social behaviour, it has been no less successful in studying experiences in the past. However, it is just as suitable for studying relationship structures (Breuer et al., 2019; Morse et al., 2021: e.g. 50, 165). Studying experiences and relationship structures is particularly important for a closer look at social behaviour, which I examined here and provides insights into creativity and the understanding of innovation. In the context of GT, however, I will take the liberty of integrating John Law's idea of 'method assemblages' (Law, 2004: 13), although I do not treat it as a second methodology next to the first but rather as a conceptual extension of it. The integration has the purpose of not losing sight of the 'hidden, small details' that are also part of the mosaic, although they initially do not seem to fit the systematisation. This is not to say that GT would not allow this, although I would like to extend the focus to the subjective, which sometimes threatens to disappear. Above all, I found inspiration in not having to talk about concrete emotions to access them. The idea is thus not to have a conversation that starts with discussing feelings since, as I mentioned in Chapter I, this caused misunderstandings. Instead, the idea was to develop a conversation full of emotions regardless of the concrete naming of personal feelings. In my material, sentences such as 'I feel sad because…' are rarely found, and instead, sentences widely occur that elaborate at length, describing a situation, followed by an emotion, sometimes abstractly, sometimes concretely. However, this is not because I attempted to stage a therapy session, as this would have been inappropriate and presumptuous, but because it is part of the work that is to be located at the places of work and beyond—in private.

Accordingly, it is not only the methods but also the researcher's gaze or approach that play a significant role in how (and what) data is collected.

Therefore, I avoided the problem of a static view by using various methods recommended by GT. In this respect, my flexibility in view and method was indispensable, not least because of the different obstacles that had to be overcome and which emerged during the research. In the work process, it became increasingly clear how important it is to critically examine the topic in addition to having a reflexive attitude towards oneself and the field. Studying the handling of innovations and the term itself is a very vague undertaking, and the fact that the results are sometimes ambiguous is not due to the 'messiness' (Law, 2004: 18) of the study but to the intricate policies of the matter itself. The mere conveying of emotional content says nothing about when it has contributed to a decision, and a very reflexive attitude towards one's emotions can filter them down to such an extent that no content worth mentioning remains.

The accompaniment of various actors from different groups and disciplines ensured a constant change of perspective. By this, I mean that the field is first seen through the eyes of the informant, and only the researcher who accompanies them can finally bring the different perspectives together and provide an overall picture.

Following John Law's proposal to move away from the classical understanding of methods toward 'method assemblages' requires 'a combination of reality detector and reality amplifier' (Law, 2004: 14). This involves discovering the respective quality of a method and applying the method against this background and with this claim. Methods are less to be seen as already given and only used as a means to an end, significant only in their self-evidence. With this in mind, Law criticises a researcher's all too common approach in his methodological approach to the field as the standard routine possibly tempts one to become redundant in applying it and not remain purposeful. This results in everyday descriptions that reproduce themselves but cannot provide any significant gain in knowledge. Instead of investigating 'the everyday', the method produces 'an everyday' with perennial answers to one's questions (Law, 2004: 45 f.). Hence, to avoid the error of one's own bias and discover the many realities out there, researchers are encouraged to develop quiet, slow, and modest methods that do not tempt us to be overly imperialistic (Law, 2004: 15).

To return to GT in combination with the method assemblages according to Law, it is important to note, especially for this work, that one's own perspective is not only challenged by constantly aligning itself with new disciplinary cultures. As I will demonstrate in the empirical part, this approach reflects the overall attitude of this research topic. Not only does the topic require me, as a researcher, to constantly be flexible in changing perspectives, but the same applies to the actors in the field who, in the course of their work, have to engage with many others who do not share the same background or perspective. Apart from following the prototype, it was about observing how many cultures situationally become one to create something together. Thus, a story of innovation and prototyping unfolds in different places, in different group constellations, and through various languages.

> So assemblage is a process of bundling, of assembling, or better of recursive self-assembling in which the elements put together are not fixed in shape, do not belong to a larger pre-given list but are constructed at least in part as they are entangled together. This means that there can be no fixed formula or general rules for determining good and bad bundles, and that (what I will now call) "method assemblage" grows out of but also creates its hinterlands which shift in shape as well as being largely tacit, unclear and impure (Law, 2004: 42).

Accordingly, stories and realities are '[n]o longer independent, prior, definite and singular as they are usually imagined in Euro-American practice. They become, instead, interactive, remade, indefinite and multiple' (Law, 2004: 122). This approach is compatible with GT in the following way: while a systematisation can be undertaken for some narratives, this is not possible for all findings. Thus, I have made it my goal to go beyond the 'pre-given lists' (which I will revisit in subchapter 2.3) and embed individual, non-sortable inferential analyses, for it is precisely the seemingly marginal that comes into effect through contextualisation. Consequently, on the one hand, realities and associated narratives develop very differently, and on the other hand, they are interconnected and refer to the same phenomena. Interestingly, however, the difference in the narrative does not depend on the person who narrates but much more on the place and time. Hence, it can be that a person speaks and feels differently at different times and in other places than previously described (Gammerl, 2012; Law, 2004). Similarly, different people may develop a similar narrative at varying times in other places. This observation is eminent because the person in space and time makes the difference. Therefore, it is again justified that the focus of the study is on something such as human emotion and what is detected as a feeling.

> Narratives and their enactments are not fixed [...]. They are negotiated and renegotiated. The fact that they are negotiable and in need of negotiation is entirely explicit. So too is the fact that those negotiations are strategic in character. The implication is that if singularity is achieved (and the extent to which this is the case is contingent and uncertain) then this is a local and momentary gathering or accomplishment, rather than something that stays in place (Law, 2004: 129).

It is precisely this demand for flexibility and fluidity, justified by vagueness, that only further substantiates the question of the emotionality of technology development, which ensures a new approach to innovation practices. All too often, these are pseudo-rationalised processes that are treated as such. In this respect, the investigation using qualitative methods makes sense and, further, an approach that does not exclusively ask about policies and structures. The connection between the question and the chosen method examines, on the one hand, the communicative forms of and around prototypes, such as translation narratives (translations) of innovation.

In what I have examined, I refer to feelings and emotions that are generally considered as such. These include, for example, anger or frustration. In addition, I draw on the statements of my interview partners who, for example, describe their gender in connection with discrimination as a feeling of oppression. Finally, so-called meta-emotions (Archer, 2000: 224), such as trust, are also part of the investigation, although I do not make an explicit distinction every time I name them. While emotions and their temporality and the above-mentioned differences may seem fragile to some, they refer to an object, relationship, or activity. Therefore, GT is able to grasp these expressions because they refer to these fields.

2.3 The Research Sites

As already indicated, I do not describe individual case studies based on their prototype examples. By contrast, I will describe and analyse the research sites visited that were relevant to the work. In the later evaluation, it will nevertheless become clear to which location and prototype I refer.

While the collected consideration of the phenomena may seem unusual, it has the advantage that they can be considered independently of their context, as they do not represent individual cases. In this respect, the emotions that occur do not represent a single case scenario per se but have recognition value for others.

Furthermore, for reasons of confidentiality and intellectual property (IP), only partial details of the prototypes will be provided. In concrete terms, this means that the descriptions of the respective prototypes are more like the published descriptions of the inventors and teams and do not reveal any technical details beyond this.

In total, I visited four different sites of invention and/or innovation, where I was able to observe several developments, and I will now discuss both the places and the prototypes in more detail. The first site is an *incubator* that is connected to a university hospital, where I studied two different teams and their invention. Secondly, I did research at a *makerspace*, which I observed as a whole complex, meaning that I was not observing particular teams. The third site was a *creative space*, which I did not observe as a geographical space like the makerspace but rather as a mental one where ideas develop amongst people. Finally, I looked at team dynamics in an established *company* that has been operating since the early 1990s. In addition, I consulted an *innovator* and private investor. His work and inventions are examined from two different angles, and describe independent impressions that are very similar to those of the others. In this section, I will also mention who or how many technological artefacts were specifically accompanied.

To ensure confidentiality, all names, those of the innovation spaces, the company, the teams, and the interviewees, were anonymised and changed.

1) The Med-Tech Incubator: Health Hub

The digital labs of the publicly funded incubator are dedicated to biomedical developments and form part of a more extensive healthcare network that is exclusively available to doctors at a specific clinic and research centre in a German city. The doctors at this site pursue the mission of translational medicine, aiming to transfer biomedical research results and inventions to patients or launch them in the healthcare market. In this respect, the incubator uses an annual call to invite doctors from one clinic to realise their ideas. After a successful application, they are given money for a certain period, at best – provided that they successfully fulfil their milestone plans – several years. In addition, they are released from their work at the clinic for this period while they continue to receive their salary.

The incubator initially provides its successful applicants with a room and the necessary equipment. Most of the time, these are ordinary office rooms that resemble co-working spaces. In addition, the doctors and later teams are advised by the incubator and external consultants. They have access to a wide range of services and can distribute orders, e.g. have preliminary prototypes developed or hire developers. Before they have to take part in so-called demo days to demonstrate their prototype to a selected public, they can participate in drama workshops to reduce their nervousness and practice performing for events. In these workshops, they are taught how to deliver the best possible performance in order to market their idea, e.g. to representatives of health insurance companies. In addition, the incubator grants them access to a broad network and interdisciplinary collaboration in the labs.

During my fieldwork in the incubator, I alternated between talking to the Head of the Incubator and Accelerator Programme, *Jan*, and an externally contracted consultant, *Felix*, who advises and discusses milestone plans with the teams during their time at the incubator. In addition, there were informal discussions with other incubator staff, and although these were not part of the evaluation, they helped to establish connections and structures. The incubator's portfolio has broadened considerably in recent years and involves some software and hardware developments. I dealt with two hardware developments in my ethnography: a sensory insole (*Feety*) and a pulse-measuring headband for anaesthesia (*Ellie*).

I accompanied the two teams for different lengths of time. I had my first contact with team 'Feety' in 2019 and maintained it until the end of this work in 2022. On the other hand, I only accompanied team 'Ellie' for a brief period, although it was still in the early stages of its development.

Team Feety: 'Feety' is an insole for postoperative patients. After a knee or hip operation, the sensory insole, inserted into the shoe, helps measure the weight of the load. In this postoperative phase of a couple of weeks, a maximum of 15 kilos of weight is allowed on the operated side, which is why the insole, which is connected to a mobile phone app, then sends a warning to the user to remind them of the load limit.

In this team, I spoke with the physician and project and team leader, *Bahar*. She developed the idea for the sensory sole together with her sister. I also spoke with *Hendrik*, her husband, who was hired as an executive officer in her project. Lastly, I talked to *Viktor*, an externally employed developer, who was hired full-time for the project. Most of the interviews with this team were with him. This is simply because, during the time of my investigations, the other team members, including the team lead, left.

Bahar and her sister applied to the accelerator programme in 2018, together with their supervisor, the chief physician at the clinic. The head doctors are part of the application process but only play a role in the background and do not actively work on the projects at the incubator. Their application was successful, and Bahar started working on the project with her sister, but the latter soon dropped out due to her medical school commitments. In the meantime, Bahar hired Viktor, a successful founder of a social media platform in Romania, who had studied at the Massachusetts Institute of Technology (MIT) and previously worked in Shanghai. He brings a lot of experience and expertise and also has had a similar idea to Bahar's, which is why he started working on the project in the first place. In 2019, Hendrik, Bahar's husband, was hired, and they worked as a team of three for a while. In 2020, the first summer of the COVID-19 pandemic, Bahar and Hendrik stopped working on the project, and Viktor remains the only active employee and developer in the project. The conversations with Jan, Felix, Bahar, and Hendrik took place in German. The conversations with Viktor and Ryan were in English.

Team Ellie: 'Ellie' is a type of headband that measures the brain's pulse waves under anaesthesia during surgery. The headband, with its sensors, thus measures the blood supply while the body is immobilised. After an anaesthetic is administered, the so-called Postoperative Cognitive Dysfunction (POCD) syndrome can occur, which is characterised by postoperative delirium and cognitive dysfunction. This dysfunction is caused by a lack of blood supply to the brain. POCD occurs relatively frequently and does not always have a lasting impact on the patient, although the undersupply can lead to a long-term problem, which means that people can no longer live independently afterwards. The headband is designed to help monitor the blood supply during surgery to improve the blood supply if necessary and thereby prevent POCD syndrome.

I spoke to Ryan, 'Ellie's' project and team leader, twice in English. He is also a physician and applied for a position at the incubator in 2019 with his supervisor and chief physician. The idea for 'Ellie' did not come from Ryan himself; he was introduced to it by his supervisor, who already had contacts with an external development company that develops engineering services for the medical sector, such as pulse measurement. In this collaboration, the chief physician and the company came up with the idea that pulse measurement could also be used in anaesthesia. Ryan was enthusiastic about the idea and developing it further in the clinic's incubator. He worked alone until mid-2021, when he hired a developer who now supports him in his daily work. With 'Ellie', too, the chief physician stays in the background, working at the clinic. Ryan and his developer, Shahaf, are now fighting several battles. Due to the inclusion of the external service provider for medical technologies, which already took place before their application to the incubator, conflicts often hinder the project work.

2) The Makerspace: *M.lab*

This makerspace is a 4,500 m² hard tech innovation hub that is home to over 70 start-ups and has more than 500 members today. The space was founded in 2015 by Christian, Filip, and Baris. The website indicates that, like many start-ups, it started with a small idea and very personal challenges. As for the founders, it involved implementing their own idea: a bicycle with an alternative drive. Through their work on the bicycle, they figured out how helpful and necessary it was to have adequate space, equipment, and a network to rely on. Therefore, they founded this makerspace after discovering the need for such infrastructure. When I visited the space for the first time in 2018, it was still in its infancy compared to its present developments and offerings.

Today, in 2023, the space disposes of an advisory and mentoring programme. One can get advice on various levels, also concerning a suitable space and the operation of large-scale equipment, such as the 3D printer or the laser cutter. Above all, one can be advised regarding one's idea and a potential start-up. There is an extensive network comprising business angels, i.e. investors that one can draw on or be referred to. The so-called academy also offers support with concrete problems that have to be overcome in prototypes and design.

I conducted interviews with Christian, one of the three founders of the lab, in German. Christian is their frontman and works full-time for the lab, whereas the others operate in the background or have other obligations. Christian also had the idea for the alternative bicycle drive, and his prototype adorns the makerspace's lobby. He is in his 40s and manages the place at present. When he enters the makerspace, people salute him. He has a little chat at the counter or with people he runs into. He usually knows the people who are currently working on some project, is in touch with the people and is open to conversations, especially when it comes to the point where inventors need feedback, advice, or contact with the network.

The makerspace usually differs from hubs or incubators as people rent a space in the lab and initially invest their private funds since they usually develop their ideas with their own money and are not sponsored by investors in the early stages of development. This approach makes their work comparatively risky in terms of financial losses. Further, the inventors do not receive any salary and usually live on their savings or work on their projects in parallel to their regular employment.

3) The Creative Space: The Believer School

The creative space *The Believer School* differs from the makerspace in that it is not a space for rent or to develop a project over a more extended period. The space offers workshops at the interface between art, technology, design, and human connection. Susan, the founder and manager of the school, wants her students to deeply engage with their daily life surroundings and to be critically minded when it comes to interactions with what they want to create. Susan's understanding of the school is to enforce its students' curiosity and creativity. The student body comprises artists, researchers, designers, and others who intend to find new topics to work on or who want to gather inspiration or a starting point for what they might call a vague idea they want to discuss. The programmes are either held for one weekend or take place for several weeks, and the participants can work alone or team up with others from the workshop. Sometimes, ideas overlap, or others are enthused by an idea that they want to work on. The workshops initially discuss in what environment people work, live, or spend most of their time. After such discussions, the instructors may ask further questions that take place on an emotional level, such as: 'What bothers you most on your way to work?' or 'When you think of your first/ last kiss, can you describe what your skin felt like?' Although these questions may appear random at first, they are intended to encourage people to think of something they connect with in their daily lives and to be able to describe such things on an emotional level. These workshops live from felt experiences that leave lasting impressions on people's lives as what comes up in the discussions or through such questions serves as a starting point for later ideas. Afterwards, people can introduce their thoughts or ideas they already have or tell others what they want to achieve during the weekend or over the next couple of weeks. Later, people are instructed in inspiring topics, tools, and maker sessions. Occasionally, the workshops are also about prototypical interventions in public spaces.

After a workshop ends, exhibitions or the first prototypes will often be further developed in another place, for example, a makerspace. Susan's creative space offers workshops on various topics. I took part in such a workshop, albeit not as a participant but as a researcher, and on two afternoons, I accompanied and observed people and their ideas and had several informal conversations. Additionally, I exchanged ideas with Susan about the programmes and prototypes that had a medical context and were meaningful to my work on a psychological-emotional level. Susan and I spoke in English.

4) The Firm: 'Hydro'

Hydro is a globally active company that has existed since the early 1990s and develops neurosurgical hydrocephalus valves. Johann, the Chief Executive Officer (CEO) and founder of the company applied for funding from the German Federal Ministry for Education and Research (Bundesministerium für Bildung und Forschung; BMBF) at that time with the idea of developing hydrocephalus valves and thus founded the company. Hydrocephalus is a pathological expansion of cerebral ventricles of the brain that are filled with cerebrospinal fluid, leading to increased pressure in the skull and displacing parts of the brain. During this displacement, vital parts of the brain can become trapped, and a valve like the one developed by *Hydro* can alleviate this.

Johann, amongst others, has developed improved technology to drain the water from the brain and used titanium for the valves, which are more suitable for insertion into the body. Today, the company, one of the biggest valve providers worldwide, is located in a middle-sized city in Germany.

I got to know Johann in a private context, in which he mentioned his firm. He immediately caught my attention, especially because he was talking about innovative medical technology. Conversely, Johann could also immediately relate to my research question and support the focus on emotions in research. After our exchange, I got his business card and an invitation to his company.

Johann and I had one on-location interview and later had phone calls and informal exchanges via e-mail. I also took part in one team meeting. Later, he put me in touch with one of his product developers, Leo, with whom I had two interviews and who explained the technology and the physics involved again. However, he was careful not to share any internal details on new technology developments, and I was provided with a great deal of material that was innocuous for the company. I later decided to use the website as a database precisely because its content is rich in emotions and values.

What Johann and Leo repeatedly mentioned in the conversations, and which the website also postulates, is the importance of listening carefully. As a product developer, communicating with patients, their families, and neurosurgeons is a unique challenge that takes up much space. This remark came up repeatedly and also underscored the general guidelines of the company's cooperation with doctors and patients. All our interviews were conducted in German.

5) The Private Investor and Innovator: Karwen

Karwen is a private investor and innovator in his late 30s. He came to Berlin ten years ago as a developer from Lebanon to work in a former start-up that quickly became a big company. After a brief period, he and two of his colleagues began to realise their own ideas and founded companies themselves. Some ideas were more successful than others. He soon quit his job at the company to devote himself to his own ideas. In the meantime, he has been able to build up a financial cushion to invest in young innovators himself. Karwen knows the pitfalls of founding a company; not everything he has founded or invested in has been successful. To date, he has founded his fifth company and invested in four others.

To not endanger the currently emerging start-up and the still very young idea, Karwen's idea cannot be elaborated on in detail. However, his effort resulted in the development of a website in the medical field that primarily serves to refer patients with specific symptoms and diseases to suitable facilities, doctors, and other personnel in the medical area.

Karwen and I already knew each other, and we got back in touch when I heard about his current idea by chance. We had several conversations, which often took place during his lunch break. These were held in a mixture of German and English.

The interview partners and venues in relation to each other:

Next to the company *Hydro*, the incubator *Health Hub* is the most established and institutionalised innovation venue. What distinguishes the two is their funding.

Even though the BMBF initially financed *Hydro*, it is now an independent company that funds itself.

On the other hand, the *Health Hub* incubator is primarily financed from two sources: firstly, from public funds from the federal and state governments and secondly, by a foundation. This results in a different dependency for the incubator, which is thus obliged to report its expenditure. In addition, there are strict guidelines on what the funds are used for and whether tenders must take place. The company is also not a public place that random people can visit as in-house innovation, and further development of products take place on-site. The incubator mainly invites doctors from the university hospital to take part, and all other employees, such as software and hardware developers or consultants, are hired externally.

The makerspace is open to anyone who wants to rent it and further develop an idea. The infrastructure is equally available to those who have a membership, whereby they are free to determine the length of time needed to complete their work. As a creative space, *The Believer School* is not a co-working or makerspace like the *M.lab* but rather a mental space for developing ideas. Although tools and limited infrastructure are also provided, these are only available for as long as a workshop lasts. The concept is not about sections of space that can be rented individually but about coming together collectively, exchanging and using synergies. What is in the foreground here is the idea of a completely independent market.

It is different for Karwen, who is looking at the needs of the market. He runs out of competition compared to these described places because he wants to start his own business, and hence, the referential link is not the place but much more the activity description. What the interlocutors all ultimately have in common is the portrayal of their ideas, motivation, and the subsequent work process or activity. These all differ in their content, although they will also have similarities in how something is accomplished.

2.4 Materials, Methods, and Analysis

This study approaches the research question ethnographically. Ethnography is designed to mainly combine qualitative methods to draw closer to so-called everyday cultures (Breidenstein et al., 2020), whereby the concept of culture is broadly applied. In the context of this work, the concept of everyday culture is applied to the research question in two ways. Firstly, the term 'everyday culture' refers to the general multiple uses of the word innovation, which creates a culture of the term itself. Secondly, the concept of culture is applied to the everyday work life that I am researching here. Specifically, it is about interaction practices, the communication that takes place, and the potential structures that result from it, which in turn lead to the development of a 'culture' negotiated by the people in the field and subsequently referred to as such (Ciaudo et al., 2021).

To get in touch with potential informants, their ideas and work, I first e-mailed them an interview request, expressing my interest and research question. I notably picked informants and institutions that develop medical technologies in an innovative context either because I came across them on the internet by explicitly looking for them or by chance. In addition, as the reflection section explains later, several interviews resulted from private encounters and referrals (snowball method) (Mannik & McGarry, 2017: 71), as simply contacting unknown institutions via e-mail often remained fruitless and mostly did not yield a response. After the voluntary informants mapped out the field, I used different qualitative-ethnographic methods to follow their ideas, prototypes, work, and working environment. I will list the methods used below and refer to the output generated.

To start with, participatory observation was the first step in my approach to the field, which yielded personal field notes, visual notes such as my drawings or sketches, and observation protocols as a reflection of my earlier notes and the photographs I took during my visits. I also used a preliminary field diary to reflect my position and relations with the interview partners. This diary became increasingly relevant when I sometimes found the innovation field inaccessible, as it provided a space to reflect on the ways of access that were not working so that these thoughts were supported to consider how access could alternatively be granted. Secondly, the first interviews with the individual institutions or interview partners who consented to participate, such as *Health Hub* or *Hydro*, involved a guideline-based approach in case the discussion would not develop freely. It was only in spontaneous group interviews that happened by chance or when the individual interviewees and I already knew each other that the conversations took place without a pre-formulated guideline. Further, sketches drawn by the interviewees outlining the prototype's genesis or work steps during the interviews were subsequently part of the generated data material. Along with conversations, I later tried to accompany them in their work (*go-alongs*) (Flick, 2018: 349). Accordingly, interviews and in-situ narratives (Amelang, 2012: 148) took place to understand the work and purviews of my informants. This method, however, is not necessarily valuable for recording dialogues or talks, which is why I often took minutes from memory. These three practices of narration differ in that interviews aim to answer concrete questions, and thus, the interviewer is primarily responsible for conducting the conversation and is more capable of guiding and directing it. In contrast, the *go-alongs* and *in-situ* protocols with their *sketches* follow the direction of the informants. Now and then, I also captured these activities as *photographs*, depending on whether or not a suitable moment arose, using the camera as a recording instrument. However, I did not want the camera to intrude, as, in addition to my presence as an observer, photo-taking might already be unnatural in the usual work habitat. Finally, I analysed websites, videos, and brochures of prototypes and products that were developed by the teams. I did this to inform myself about earlier narrations of an idea or to observe how the storyline about a prototype developed over time, as well as in comparison to the interviews that were conducted. An overview of the methods and types of data generated is provided in *Table 1* below.

Table 1: Overview of Methods and Types of Generated Data

Method	Type of generated data
Participatory observation	Field notes: subjective notes and spontaneous thoughts on actors, sites, and theory or methodology (= descriptive notes); drawings; observation protocols (= reflective notes) Field diary: reflected own position and feelings in the field and dyadic relations with interviewees (sometimes recorded and transcribed afterwards); sometimes taken from memory if necessary Photographs
Following in-situ narratives: work-alongs	Recordings if possible → transcripts Minutes that were taken from memory Photographs
Taking semi-structured interviews	Recordings → transcripts Sketches obtained from interview partners
Comparative method: document analysis (websites, videos, brochures) as comparative material to interview content	Essence derived from websites and brochures Discrepancies in story telling

In this section, I discuss the definition of the material, the situation in which it was gathered, the direction of the analysis with the theoretical delimitation of the question, the type of coding, and the codes. Finally, I examine the compilation of the results and the respective interpretations. As previously mentioned, this study and its structure of the text analysis find inspiration in the approach suggested by GT. This qualitative analysis explores textual documents such as interview transcripts, websites, or brochures.

Moreover, the type of analysis helps to evaluate the data as objectively as possible. This means that, despite existing subjectivity, the researcher is encouraged through various work steps to reflect on the research question, the interview questions, and the analysis of the transcript and other text documents and to regularly check whether the evaluation is meaningful. To do this, I describe the various steps

I took during the text analysis below. Qualitative text analysis is to be understood as a method of analysis alongside those already mentioned and is not in competition with ethnography as such. Instead, it is intended to support applied ethnography and to help find an answer to the research question.

First, several questions need clarification before analysing the material. As a first step, all interview transcripts, as well as other documents such as the websites of the innovation locations, interview guidelines, drawings as made during the mentioned case studies, and photos, are defined as the material of this work. Secondly, it is important to mention that the participation of the interview partners was voluntary, i.e. without any compensation agreements and that they were either contacted directly by me or via acquainted interview partners. The interviews with the founders, innovation hub leaders and team members were initially conducted using guidelines. The formal characteristics of the material (third step) can be described as the fact that the interviews, which were recorded with a dictaphone, were transcribed on the computer with an artificial intelligence (AI) transcription aid and transcription software and then imported into the analysis software for coding the material. The analysis aimed to describe the object of the research dealt with in the interviews and the documents – the emotional impact on the technological development process – and to analyse it contrastively between the cases.

The fourth step clarifies the direction of this book's analysis, whereby the analysis of the material is interrelated between the actors involved and the artefact. Through the interviews and statements of the participants, comments about the structures, contents, procedures, processes, and effects of emotions on innovation will be documented. The theory-based differentiation of this question as the fifth step results from the fact that psychological and sociological studies on emotions and their impact on interpersonal relationships are available. However, it is now of interest whether these factors also have a similar effect on technological developments or to what extent evoked emotions have a relevant impact on development paths and whether the prototypes, in turn, can equally trigger something in their observers. During the code analysis of the material, the first step was to code in vivo. In vivo coding is part of the open coding approach and involves a more rigorous analysis of words and sentences with codes that emerged from the discussions with my interview partners. In the second step, I constructed codes myself if no suitable wording was provided by the interviewees. However, this applied more to general state descriptions, such as 'evaluation' as a cluster code differentiated into 'government aid'. This – as an example – was when an interviewee explained to me that they were financed publicly, and I knew that, therefore, they were dependent on successful out-licenced products to justify the funding. Alternatively, I coded sentences accordingly to further differentiate what was said, which resulted, for example, in an in-vivo code such as 'problem' and the further differentiation in my construction as 'gender relations'. This was done to provide an overview of the material and, in the spirit of method assemblages, not to examine the material in a biased way and limited to the theoretical framework. As a result, this also means that in places where it was impossible to group codes into clusters, the 'open codes' were left and marked with a specific colour. This is reflected in the material in individual analyses that only apply to a specific case that is nevertheless meaningful and significant. For example, certain decisions become visible based on personal history and ethnic origins (see, e.g. subchapter 7.4 on Karwen, who decides against the production of war technology or its field of application because of his experiences in his home country).

Precisely because the field of innovations can be both challenging and surprising, it is essential to maintain this potential, especially in terms of its emotionalisation, and not to rigidify it through previously determined units of analysis. The codes that emerged from the open analysis formed the first frame of reference, which was later aligned with the theory or used to build further theoretical concepts that I did not consider beforehand. This also ensures that the theory is expanded or replaced in case of doubt if inconsistencies emerge or assumptions from the theory do not fit the material. By comparing the open codes with the theory, theory-based codes could now be added, through which a second analysis took place. This occasionally resulted in theory-guided codes replacing already existing constructed codes with more suitable terms. However, this was only the case when the direction of the analysis pointed to the research question.

In contrast, the codes continued to coexist if they took different paths. It was thus essential to attach the code to a targeted definition so that no duplications or inadvertent changes in interpretation occurred in the repeated coding process. The codes were then thematically ordered and clustered so that the statement and weight could be read off in the subsequent analysis. Especially concerning the interpretation in connection with the comparison of the cases, the code cluster, meaning the superordinate theory line, was indispensable and particularly valuable. In this way, the weight of a code, i.e. the frequency of an occurrence, could be seen immediately, which was especially helpful when reading certain emotional expressions of the interviewees. The cluster designations reappear as superordinate sections in the empirical parts. The deductive derivation from the theory-led coding level provides the framework for the two theory chapters (III and IV), which are an extension of each other regarding their content. Imagination, experience, and emotion are thus equally the content-related frame of reference for the described innovation structures (geographical and conceptual) and the resulting evaluation patterns. Non-clusterable codes, on the other hand, remained as individual colour codes, as I have already described, and were embedded in the narratives as individual phenomena that stand for themselves.

2.5 Reflexions During and After the Fieldwork

As already indicated in the introduction, specific problems arose due to the research field and the timing of the research. In the following, I will describe the inherent issues in the field and refer to problems in the empirical survey, which mostly took place during the COVID-19 pandemic. The below descriptions are mainly based on my field diary notes.

1) Confidentiality and Discretion: Finding Adequate Interlocuters

As described, it is highly challenging to research a complex field such as innovation. Already at the beginning of the empirical study, when I approached potential institutions, guarantors, and patrons with the request to talk to them and research them, I mostly received refusals. Out of 28 interview requests, I initially received four acceptances, and it was even extremely difficult to get in touch with institutions that normally cooperate with the university and are publicly funded. Innovation sites are black boxes. Both their structures and what is to be developed are kept secret in the fragile stage of not-yet-being-finished. Even if I managed to get in touch with a person, an institution such as an innovation hub, or visit such a place, the encounter often ended after a first conversation. If I had the feeling of having a foot in the door and being in contact with several people, it could happen that I received an urgent e-mail prohibiting me from either conducting an arranged interview or a participant observation. This was frustrating, and I often felt like I was treading water or not progressing with my research. Being a researcher often gave me the status of an outsider who should not know too much, should not look too deeply, and should not know the structures intimately. I often got the impression that empirical research was misunderstood, and I had to explain the need to understand the whole fabrication process to the same person several times. However, even these efforts did not grant me access, and hence, I often gained access to the aforementioned informants through private contacts who referred me to them. This was the case, for example, with *Hydro*, whose CEO I met at a wedding and who invited me to visit his company. In the end, it was through his recommendation that I could talk to other employees.

Interestingly, the CEO was much more willing to talk than his employees, who feared that internal company information would leak out through me. Through repeated coordination, we were able to agree on what content could be shared that would be meaningful for me and would not harm the company. It was similar to Karwen, whom I have known for many years. By chance, we met again, and he was developing an app in the health sector at that very time. We quickly started talking about it, whereupon I asked him to conduct an interview with me for this work, which resulted in several conversations. I am convinced that the pre-existing acquaintance was a door opener and helped Karwen talk to me more openly than any other interview partner about creating ideas, innovating, and monetarily investing in ideas. Susan, the founder of *The Believer School*, was the only contact who responded positively to an e-mail interview request and willingly told me about herself, her work, the school, and the workshops. The conversations with her were long and rich in information. Susan struck me as an extraordinary interlocutor, alive with ideas. Her creativity was palpable in what she shared. It was not just a description of an activity, as the way she described it was no less meaningful than the activity itself. Although these conversations sometimes led in different directions than I proposed – as seen in the material – what she said is particularly meaningful.

The situation differed from that of *Health Hub*, where access was difficult even though my supervisor paved the way for the first contact. Later, access was granted through another project I worked on, so I was able to use the access and the material in a double way. The initial contact with the hub's head was optimistic at first, and he put me in touch with a team. Afterwards, however, it quickly became clear that the contact with me should not be too close, and I wrote numerous e-mails to which I received no reply. Attempts at mediation often failed, and once, I was actively discouraged from further interviewing a team. This experience was frustrating. In addition, the hesitant impression was further reinforced even at a later stage. However, access already existed, and my research interest was known; I had to submit a written application and again set out in writing what my project entailed, what questions I had, and who exactly I wanted to talk to. After a clarifying conversation, I could be placed with another team, but no more guarantors emerged who were willing to talk to me. It became clear at that moment, however, that it was a public incubator affiliated with the university and, because of the partial public funding, was also subject to more stringent accountability.

2) Finding a Way to Talk About Emotions

The second difficulty I faced involved 'tracking down' emotions. As also evident in the empirical data, emotions hardly play a role in the context of knowledge production, and if they do, then only in a reflected meditated way.

The first problem related to my research question was the apparent lack of definition of the emotions I asked about. The informants wanted me to specifically name the emotions they were supposed to discuss, to which I did not comply. The open-ended nature of the question was intentional precisely because I did not want to prejudge the nature or definition of the emotions. Moreover, when we did get to talk about it, it quickly became apparent that innovators and their team members do so in a rather meditating way. By this, I mean they were talking about past emotions, by then reflections, that were already processed and, therefore, not pure or *raw* in terms of how they were experienced. Margaret Archer would call these 'second-order emotions' (Archer, 2000), and she refers to this process of reflection as an inner dialogue that comments on the emotional state. These emotions are indeed constructive; however, their interpretational status is determined as it is already clarified for the individual who feels them. Coming to so-called first-order emotions was much harder as it required getting the interviewees into the mood to share emotions that they had not already reflected on. The purpose of these emotions is that their emotional content is not yet being guided, and consequently, there is more space for what people share when they need to explain situations to me. They do not shorten a narrative as they have already decided on the emotional stance, and thus, there is more material I could work with in the ethnography when coding the interviews.

Occasionally, however, it was helpful to be with the group during the observations and also the on-site interviews. Arguments, even those behind closed doors, could be observed more easily, and what is more, they could be discussed in the subsequent interview(s). These situations were so evidently emotional—just like the noisy slamming of a door—that I did not need to provide any further explanation about emotions to talk about them.

3) The Ambivalence of Talking

Another problem was keeping in touch with the people who were already willing to be interviewed. Interestingly, I found an *ambivalence* in 'explaining innovation'. For the aforementioned reason of secrecy, it turned out that my interviewees felt slightly afraid of talking 'too much', and they frequently interrupted an interview to tell me something that they did not want to be mentioned in my work. In such instances, they felt the need to talk about something but did not want it to be officially shared or opened up to the public and thus, they would pause the recording or tell me that I should not use specific information for the study. Some guarantors felt relieved by their participation in two different ways. On the one hand, they enjoyed talking to me and telling me secrets or inside information, while on the other hand, they also felt at ease once the interview was over. I reckoned that once an interview ended, the contact with a team leader or team member would terminate, and it often required a continuous effort to keep in touch with the teams, especially during the COVID-19 pandemic-related contact restrictions when it was not possible to meet in person.

Sometimes, I would not hear back from teams for months, and then, all of a sudden, they would invite me for an interview. In other instances, I needed to write three times before I would receive an answer, even though the respondents would tell me how much they enjoyed talking to me. However, the extra exertion was worth it, as it paid out every time, and I was also pleased to at least be able to keep in touch with this handful of teams I managed to convince to participate in this study.

4) Following the Prototype During a Pandemic

Another challenge in the context of keeping in touch with my guarantors was to then follow their materialised ideas, aka prototypes. The initial idea was to observe at least one team so closely that I could follow their idea from the very beginning to the final prototype. However, as the teams were often not working in their workspaces but from home, and not even as teams together in one room but online via video telephony, it was difficult to observe their work on the prototype. The work thus took place online or involved only one person working on the prototype at home. Workspaces were closed down; people were isolated or were also not interested in talking to me in the tough times during the pandemic when they had to cope with a great deal of stress. However, ultimately, it was the federal isolation regulations that separated the teams from each other, as well as me from the teams. At the height of the pandemic and the contact restriction regulations, conducting ethnography in the planned form was out of the question. As a result, I then conducted telephone calls or even conversations via video telephony to conduct interviews and also to keep in touch, as I had described before. Especially at the time when work structures and habits changed and the professional retreated into the private sphere, there were difficult conditions that had to be overcome each time. The research space suddenly became digital, which no one could previously have expected, and thus, an attempt was made to measure voice intonation and compare it with the content of what was said. However, the method did not go as hoped either because the desired amount of data did not materialise. For this to have been successful, it would have made sense if the subjects had kept some type of diary and recorded themselves without me being present. In two cases, this was successful via voice message.

In addition, I asked my interviewees to make sketches during the online interviews and to transmit them via the screen, similar to a virtual blackboard, which I could see during the interview. The virtual representation worked better, although some of the sketches cannot be represented in this work due to the stricter anonymisation agreement.

Theory – Thinking, Feeling, and Acting in the Moral Economy

The initial segment of this study probes two theoretical dimensions, with the first addressing the genesis of an idea. In this context, the creative impetus of imagination is drawn from everyday experiences, as elucidated by Franz Brentano's pre-phenomenological approach. It is frequently demonstrated that these experiences manifest as challenges, prompting the observer into action and invoking a mode of activity. Experiences are emotionally perceived through imprints, shaped by the observer's own realm of experience. Building upon the pragmatists' insights of William James and John Dewey, I explore how pragmatism and conscious experience facilitate the discovery of multiplicity in idea development, sparking cognitive processes. This multiplicity, coupled with the confrontation with an object, elicits emotions that, in turn, give rise to value judgements. The interplay between pre-phenomenological and pragmatistic perspectives on idea formation informs the second aspect of innovation.

Various (infra-)structures, i.e. whether dictated by political legislation prescribing specific types of innovation through guidelines or emerging from self-determined locales and resulting cultures, elucidate the ways in which innovation is conceptualised, treated, and enacted. We observe the emergence of narratives and explanatory patterns, which, in turn, shape society's patterns of value creation. The exploration of idea formation, transitioning from an individual to a collective engaged in a shared vision, along with the structures imposed by innovation, leads us to what ultimately constitutes a moral economy around an artefact. The accumulation of knowledge in this work primarily centres on feeling with the object.

III. From Problem to Possibility
Imagination, Experience, and Emotion

Innovation advertisements, campaigns, and images display glowing light bulbs, dynamic arrows, and networks that connect related terms. One sees Michelangelo's 'The Creation of Adam'-like images in which a human hand approaches a robotic hand—sparking what between them, a Promethean spark? A network is unfolding; somewhere, there is the term 'AI', and elsewhere, there are strings of numbers of zeros and ones. There are icons of cogwheels, brains, and puzzle pieces. (Book) titles such as *From the Idea to Market Success* (Bundesministerium für Wirtschaft und Energie, 2020), *The Creative Mind* (Boden, 2004), or *Where Good Ideas Come From* (Johnson, 2010) grace the scene.

No matter what one is ultimately confronted with, the glowing light bulb, the robotic hand or simply the word *idea*, the central focus persists: *creativity and ideas* take centre stage. It is the sudden idea, the envisioning of a future that enhances a current situation, the work, or the intricate thought—the puzzle piece—that presently leads to the refinement of an idea, something novel that humanity has not witnessed before, perhaps similar but distinctly unique.

Founders, inventors, and hackers often recount an idea that struck them suddenly, perhaps during sleep, linked to a problem encountered in everyday life. Their idea serves as a realm of possibilities, envisioning a better world. Imagination should precisely be the starting point for this chapter and the emergence of an idea because, even if the aforementioned images aim to promote innovation as an act of creativity, they underscore one crucial aspect: imagination and creativity are the realms of possibilities for any innovation and, consequently, are highly emotionally charged. The term 'imagination' is subsequently followed by 'experience' and 'emotion'. These three terms constitute the theoretical foundation for delineating the emotional aspects of innovation processes. They run parallel to the pragmatist approach of thinking, feeling, and acting, guiding the reader through these interrelations and their significance. These terms serve as the conceptual tools throughout this work, reappearing in subsequent sections related to prototyping and innovation developments.

3.1 Imagining Possibilities

> What would an inventor or innovator be without his idea? After all, the idea is what innovation is all about. An idea [can be a dream of a better world], and it somehow carries the character of utopia. In general, the idea always refers to a "could-be".
> *(Interview from 06/04/2020, Christian, Founder of M.lab, own translation of the German transcript)*

Imagining possibilities assumes developing one's thoughts into hypotheses that become the expression of a state of possibility in the past, present, or future. This signifies a human act of creativity to imagine a life, an interaction, or a state of affairs and, therefore, differs from the actual situation. This manner of contemplating a departure from reality can be seen as an ambition, particularly when an idea is involved. The capacity to imagine, i.e. how we conceptualise what and how, also results from experiences, which are inherently intertwined with emotions. These are cascading reactions directly linked to one another. Our imagination centres around an object evoked by external or internal stimuli (Brentano, 2015). Based on experience, our engagement and interaction with the world take on emotional dimensions, meaning a motive force imbued with feelings that stir us and thus shape our mindset in the world, our stance towards something, and our relationship with the world. This can subsequently lead to actions and reactions, especially if our idea is destined for further development.

Imagining an idea thus means a distinct form of a hypothesis. Unlike in the past, speculation about the present and future encapsulates the potential for a reality collectively experienced. It is precisely this allure of the future that often leads to romanticising proposed solutions for past and present predicaments. Speculative research, in particular, has critically examined questions surrounding social futures and their logic and rationalities over the past five to ten years (Wilkie et al., 2017). In the temporal process, the prolonging present itself inhabits the future, an ever-continuing state (Wilkie et al., 2017: 2). However, this seemingly never-ending future does not replace the present but merely adapts the same logic and rationalities of the present. Consequently, the future is strategically planned and calculated based on present uncertainties and fears, losing its imaginative essence and remaining devoid of vision (Wilkie et al., 2017: 2). Often assessed through the lens of the past, the future loses its potential for change. If the future becomes a space for shaping past and present problems, it loses its modern ideal of progress, and the hypothesis then no longer expresses a possibility but degenerates into a description of a state. Here, the anxieties and uncertainties of modern times become evident as 'matters of concern' (Latour, 2004), warranting serious consideration. However, as a glimpse into the future, they are only intermittently usable, provided that uncertainty does not spiral out of control, stifling any potential for change (Adam et al., 2000). It turns out that speculation and contemplation of possibilities often clash with expectations and the desire for security, explaining the frequent absence of a bold perspective.

Particularly for innovating, being stuck in contemporary rationalities and logic can be counterproductive, and innovation hubs, prototyping labs, and incubators reach their limits as a result. The crux lies in the fact that what is deemed 'new' does not necessarily equate to 'innovative', and the overly new faces acceptance difficulties among its potential users (Chapter IV). The definition of what is considered new remains a matter of negotiation. Nevertheless, to speculate at all, to create a vision, and to be inventive requires imagination. In this context, imagination is discussed philosophically and psychologically, especially by the pre-phenomenologist understanding of it as a mental force.

3.1.1 Imagination as a Mental Force

A mental force means that a moving force is set in motion (Brentano, 2015). In his work 'Psychology from an Empirical Standpoint', Franz Brentano precisely refers to the imagination as the starting point to develop a feeling of something outside oneself. His theory explores the *activity* and *creativity* of the mind and its ideas, conceptualising feelings as mental phenomena endowed with intentionality akin to *imagination* and *judgement*.

Although the term 'intentionality' remains somewhat undeveloped in Brentano's work and subsequently presents many difficulties, his thoughts on the subject have several intriguing and ground-breaking perspectives for his time. He views (psychological) phenomena as objects of inner perception and states that they are ideas (Ger. *Vorstellungen*) or at least founded on these. Brentano makes a clear distinction between mental[1] and physiological phenomena, categorising physiological phenomena as *objects* perceived through the senses, such as tones, colours, and tastes. These phenomena arise in the present through immediate conscious awareness during sensory experiences like seeing, smelling, hearing, or tasting.

In contrast, mental phenomena encompass acts of imagination that involve objects, such as thinking, feeling, fantasising, or dreaming. Thus, the act of listening to a tone, seeing a colour, passing judgements, and experiencing emotions like sadness, love, hate, or desire falls under mental phenomena. Initially, this awareness occurs subordinately.

However, Brentano's distinction already highlights the interconnection of experience (see subchapter 3.2), the capacity to imagine, and consequently, our emotional responses to the first two (subchapter 3.3). In summary, the realms of imag-

---

1   In the literature, this is sometimes also called a psychological phenomenon. However, to differentiate more easily while reading, the term *mental* is used here.

ination, experience, and emotions prove challenging to disentangle from one another (Beaney, 2005: e.g. 60 f.).

Thus, the inner perception (Ger. *innere Wahrnehmung*) serves as a prerequisite for every mental phenomenon. Whenever we engage in imagination, it occurs through the capacity of our imagination—often referred to as 'the inner eye'. Importantly, this imaginative process encompasses physiological phenomena, sometimes acting as the initial inspiration for subsequent imaginings. Brentano directs attention to the consciousness of a thing or a person. As elucidated, intentionality, as he describes it, signifies that a person's mental state pertains to something—an object, for example.

Consciousness is the mental faculty that allows awareness of something, thereby bringing it into existence. In this context, the reality of the idea does not hinge on its existence in the 'real world' (e.g. talking animals in dreams). The imagination of the thing (like a talking horse) renders it 'real', allowing one to contemplate things, whether real or unreal, proximate or distant. This mental capacity to be directed towards something beyond the mind, be it real or imagined, is known as intentionality.[2]

A physiological phenomenon, an experience derived from the senses (referred to as first order in the previous explanation) in the tangible world—such as an event—elicits a reaction. This reaction to an object or person may transform into a mental phenomenon. Consequently, the original experience becomes a past event. However, one's memory and the associated emotions from that past experience reside in inner perception, persisting in the present. Therefore, by recalling an event from the past, one can evoke the corresponding emotions in the present. This memory not only establishes a connection between the past and the present but also constructs a bridge from past thoughts and emotions to those of the present. Thus, the physiological phenomenon transitions from the first order to the mental one, as one cannot summon emotions through imagined scenarios without the prerequisite of a past event. 'But if physical phenomena have more immediacy in consciousness than mental acts, Brentano is at pains to point out that mental acts have a superior degree of "reality". Like all primary objects, physical objects exist in consciousness only as parts of the mental acts that contain them' (Fancher, 1977: 208).

Further, this consciousness involves a dynamic relationship between the subject and object, constituting an emotional process (Ger. *Bewusstseinsprozess*). Consequently, intentionality encompasses the direction of emotions towards something. This element gains significance as, per Franz Brentano's concept, emotions become

---

2   Although the interchangeable use of the words 'real' and 'true' in Brentano's language can be confusing, this is a compelling point in his theory as now, it indeed does not matter if something is real.

an external manifestation, serving as a connective link between the inner (my consciousness) and the outer (the world). As a result, the process of judgement ensues.

This intentionality, i.e. the capacity for imagining something, enables us to represent the world. However, what modes of representation are there? It is crucial to discern mental phenomena that provide descriptions of the world, with perceptions aligning with the real world. Desires and intentions may not necessarily align with reality but reveal how we aspire to construct it (and thus how we arrive at our judgements). Consequently, they can serve as motivation to effect changes in the world. This motivation, rooted in emotions, acts as a driving force, prompting societal engagement, political involvement, and participation in various ways (Marres, 2012). In more technical terms, emotions propel and activate individuals to either maintain a certain status quo or initiate actions that alter it (see 3.3: 'Emotions Constituting the Technological Artefact'). In both cases, it involves an activity that pertains to both the inner self and the external world. In essence, there exists a connection between subject and object, where something prompts an individual to act. For instance, a psychological phenomenon—an inner perception (idea)—elicits emotions *within me, motivating me* to take action. Consequently, *this idea has an impact on me*, generating a reciprocal reaction to the world. This dynamic can be conceptualised as feedback effects, reflecting a continuous oscillation between the inner and outer.

3.1.2 Creativity as an Imaginative Act

In scholarly discussions, the act of imagination is frequently interpreted as either constructing an alternative reality (e.g. Beaney, 2005; Byrne, 2005; Roese & Olson, 1995) or as a human faculty generating individual knowledge (Harris & Rapport, 2015: xiii). The former, as discussed in subchapter 3.1.1, involves a response to something—the status quo—that has been encountered. The latter implies a broader understanding, where it remains ambiguous whether one is reacting to the environment or independently creating an idea or knowledge. 'The imagination is a common practice, something to which human beings attend whenever they make sense of their environments and situate their life projects in these environments' (Harris & Rapport, 2015: xiii). Creativity can result from imagining a situation differently through what is learned from the environment. However, as noted by Berys Gaut (2003), not every act of imagination necessarily constitutes a creative endeavour, a perspective that appears valid. We are thus assuming an act of creativity that arises in the confrontation with the experience of reality. In this respect, as described in subchapter 3.1.1 by Brentano and the concept of intentionality, these are acts of experience that develop creativity themselves. For example, next to Franz Brentano, Berys Gaut and Michael Beaney

[have] admitted that a creative act must be intentional, which presumably involves *thinking* of something. If, at the same time, that creative act *originates* something, then since that something did not exist prior to the creative act, the thinking that was involved must have been a thinking without commitment to its existence (at the time it was thought of) [see 3.1.1: Imagination as *a Mental Force*]. So, by Gaut's own definition, "imagining" must have been involved. Even if imagination does not imply creativity, then, *creativity may imply imagination* (Beaney, 2005: 195).

Hence, I aim to delve into creativity as the manifestation of an imaginative act and provide a more specific exploration of creative thinking and 'how people invent new instances', as articulated by Ruth Byrne (Byrne, 2005: 190). Byrne distinguishes three ways in which individuals engage in counterfactual imagination, outlining three categories of creative thought: inventing categories, concept combination, and insight. Concerning the first aspect, 'inventing categories', she states: 'They may think about a category by keeping in mind some possibilities. They may think of possibilities corresponding to true instances of the category, and they may not think about possibilities that are false (Byrne, 2005: 191).' Her analysis emphasises that the imaginative process builds upon known categories, where an existing idea is renewed through 'additions.' It is not a true process of invention but an expansion of existing concepts through the incorporation of other categories. Moving on to the second category, she explores how people invent new ideas through 'concept combination' (instead of categories). 'People can understand and produce new combinations that they have never heard [of] before. New concept combinations regularly enter into everyday language use. New knowledge may emerge from the interaction of existing ideas (Byrne, 2005: 193).' This form of creativity involves the ability to combine different concepts in individually meaningful ways, focusing on word composition based on the concepts. Despite potential variations in interpretation, it underscores the certainty of interpretation in the context of the ability to combine. Byrne highlights the reliability of human imagination and interpretive ability, drawing on prior knowledge to generate something new. This underscores the interplay between experience or familiarity and innovative development.

The third form described by Byrne is 'insight,' referring to what others might label a spontaneous idea—an idea that emerges in the moment, unpreceded by contemplation, reflection, or fantasising. This phenomenon occurs without predetermined rules or regulations.

Additionally, Byrne presents a series of experiments highlighting the constraints on creativity when individuals perceive certain elements as having 'fixed functionality' (Byrne, 2005: 194). This concept refers to situations where the ontology of an artefact is perceived as unalterable and consistently serves its original purpose. 'People often add mistaken assumptions to their mental representation when they try to solve [...] problems. They can solve them when they are primed to think about alternatives' (Byrne, 2005: 195).

> To be honest, I've never considered whether imagination is important. But when I think about it now, I guess without imagining something, it's just or probably not possible. So, it's just not possible that I want to invent something and don't think about it beforehand. Something always happens beforehand, then I think about it, and then I usually know what I want to do. Or, also in another way: sometimes, in a conversation or something, I suddenly think, wow, that doesn't exist yet. We have to do that! It's a sudden idea that just comes to me. Of course, I didn't think about it for so long, but the idea is still there. It's also fun, somehow. *(Interview from 18/06/2021, Karwen, Private Investor and Innovator, own translation of the German transcript)*

In this context, distinguishing between passive and active creativity proves meaningful. Passive creativity, as Beaney notes, involves an idea that 'pops into one's head' (Beaney, 2005: 196). 'Active creativity, on the other hand, involves a deliberate process of trying out different approaches, ideas or solutions until the right one is found (Beaney, 2005: 196).' This distinction aligns with Ruth Byrne's classification, where the first two are considered active and the last as passive acts. Both experiences are recognisable. Passive creativity represents a spontaneous and uncontrolled event reacting to the environment. As previously discussed, this may not be a conscious thinking act but an unconscious event manifesting itself in a reaction. On the other hand, active creativity involves a purposeful and controlled mental process aimed at finding a solution to a problem—a conscious and goal-oriented search. Both processes are evident in practice (see Chapter V) and are apt for describing the genesis of an idea and subsequent prototype development.

In the realm of Science and Technology Studies (STS), the significance of imagining futures or future imaginaries is prominent. Scholars such as Sheila Jasanoff, Sang-Hyun Kim, and Steven Hilgartner have highlighted the influence of the rich visual heritage of science fiction in the development of technology and innovations (Hilgartner et al., 2015; Jasanoff & Kim, 2015). Notably, a somewhat staggered development of technological ideas can be observed, often drawing inspiration from literary or cinematic sources. Examples range from 'Frankenstein' (Shelley, 1897), 'Brave New World' (Huxley, 1932), and 'Klara and the Sun' (Ishiguro, 2021) to widely watched films and series like 'Star Trek' (Roddenberry, 1966), 'RoboCop' (Miner, 1987), 'Minority Report' (Spielberg, 2022), and 'Black Mirror' (Brooker, 2011). In this dynamic, science fiction transitions into science fact. 'Belying the label "science fiction", however, works in this genre are also fabulations of social worlds, both utopic and dystopic (Jasanoff & Kim, 2015: 1).' These imaginative utopian and dystopian reference reality and the present while aspiring to be avant-garde, often describing existing realities transformed to varying degrees, incorporating modern and, consequently, future visions. They seldom emerge entirely novel but rather depict an existing reality, transforming it to a discernible extent that incorporates contemporary perspectives and, consequently, future visions. The depicted objects frequently represent advancements of pre-existing categories or concepts, as discussed above. In doing so, they chart out potential opportunities that could capture the attention of the tech industry down the line. Nevertheless, these ideas and (re)presentations encapsulate entire societies and cultures that relate to reality and, thus, to a world of experience. In this context, the inception of ideas and inventions finds its roots in imagination and, consequently, everyday experience. It is through this imaginative form that such a 'bringing into being' arises, inciting emotions within oneself and others. Whether explored in the works of Franz Brentano or expounded upon by subsequent proponents of his phenomenological theory like Edmund Husserl, Maurice Merleau-Ponty, or Martin Heidegger, the intricate relationship with the world, whether real or surreal, finds expression through these channels.

Primarily, the focus lies on the 'idea' as empiricism or an extra-mental scenario. This entails an exploration of experience, as delineated in philosophy—a conscious engagement with the external world, characterised by interactions that give rise to feelings. In a more specific sense, these interactions are often construed as structured contexts of experience.

> **C:** Mhm, you are always confronted with something. With us, it's usually like this: people come here and already have a mental picture [Vorstellung] of what they want to do. Then, they need tools. [...]
> **I:** Let's go back to their mental picture. Can you describe in more detail what you mean by that? Are these already finished ideas, or how do I best envision them?
> **C:** Yes, not necessarily finished ideas, I think. From what I hear, they have concepts about what their thing should or should not be able to do. That's clear. You know, that "Swiss army knife" ["eierlegende Wollmilchsau"] thing. Something super great, perfect. *(Interview from 06/02/2020, Christian, Founder of M.lab, own translation of the German transcript)*

Hence, the ensuing subchapter diverges from experiences, referencing the imagination, as elucidated earlier, as a reality in the process of becoming. Subsequently, it enquires about problems and possibilities through a pragmatist lens, more precisely exploring the space of possibilities that the imagination unveils concerning the future. Thereafter (subchapter 3.3), attention is directed to the sociality with objects, elucidating how emotions play a role in constituting a technological artefact through value judgements. The concluding subchapter examines how an individual idea evolves into a collective idea, representing a form of knowledge accumulation in the realms of innovation theory and industry. Diverse economies sustain the iterative loops within the disciplines involved in the development and design process. Throughout these segments, expressions of feeling are significant as internal responses to the external environment.

Figure 1: Experience-Imagination Interrelations

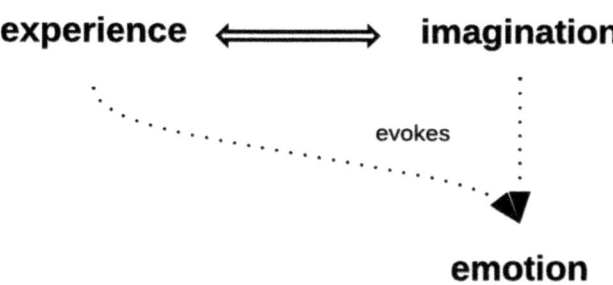

While there are parallels between Franz Brentano, a precursor of phenomenology, and the pragmatist representatives William James and John Dewey, phenomenologists would emphasise that pragmatism starts with experience as a given, while phenomenology does not require any experience to arrive at a world conception. Brentano's work already indicates a focus on closely examining phenomena as appearances. However, the assumption that this 'given' accurately represents things as they truly exist in the world is not inherently tied to it, although implicit pragmatism will later influence Edmund Husserl's phenomenology.[3] Nevertheless, in this work, the two philosophical schools are not presented as rivals. The upcoming chapters will consider (conscious) experience as both the experience of interaction and the experience of one's own imagination as a mental form from which creativity can emerge. In this context, experience does not necessarily have to be limited to what is encountered through interaction; instead, becoming aware of the imagination is also regarded as a form of experience.

---

3   Edmund Husserl is not relevant in this work. Yet, to understand the similarities among the two philosophical schools, phenomenology and pragmatism, a reference can be made to the method of *epoché* as a phenomenological reduction. In this approach, the implicit pragmatism of Husserl becomes clear. The concepts do not need to be separated either for this work or in general – especially not when at times, it is only a matter of terminological preferences.

3.2 Experiencing Daily Life

> The ideas come from everyday life. [...] They [the physicians] might see a problem in their daily clinic routine, and then they want to change something. *(Interview from 13/07/2020, Felix, Consultant at Health Hub, own translation of the German transcript)*

Experience assumes a challenging position within the sciences, given its initially subjective, individual nature and limited verifiability. In contrast, the verification of the truthfulness of such experiences may not be a common concern unless someone perceives a direct impact. Scientifically, it isn't regarded as objective knowledge in a realistic sense, lacking a claim to scientific objectivity. Pursuing a singular truth through experience, seen as a subjective perception, is deemed too sporadic to make reliable statements about reality or the nature of things. Modern philosophy has extensively debated the controversial nature of experience, particularly in relation to emotions, as emphasised by pre-phenomenologists and psychologists. Conversely, experiential knowledge becomes valuable when decisions are grounded in scientifically robust foundations (Wynne, 1998). This draws a dual perspective through experiential knowledge, aiming to complement theoretical understanding.

> Well, we are constantly making experiences, whether in the hospital or here at Health Hub. And then, not only with the product but also always in the team, with the people; probably everything flows into the work somehow. I just know that good experiences are pleasant, but mostly, the bad ones help us innovate. Because then we know what we must change and what we still need to do. In brief, bad experiences are the ones that make us think and are easier to sell as a result.
> *(Interview from 30/01/2020, Bahar, Physician & Innovator at Health Hub, own translation of the German transcript)*

From an empirical standpoint, experiences are understood as all that *has been*, provided it has been consciously felt. Our memory influences *a* given present and, therefore, *a* possible future. Experiences are identity-forming, and their collection forms the basis of how and what we feel as they provide information about an ultimate, collective as well as individual reality through a network-like interplay of situational and momentary interactions. Experiences are resonant interactions of human beings with their world. We retain those that we are not necessarily aware of now (e.g. trauma) but which can nevertheless be evoked in our memory. These memories take place individually or collectively, are part of our identity, and have an immense impact on our emotions and how we value an object (Husserl & Ströker, 2012). Actively recalled memories forge an unbreakable link between our past and present and potentially influence our future—a phenomenon akin to sensing our past, echoing Edward Shils' perspective (Shils, 1981: e.g. 52).

However, the establishment of the sociology of science in the mid-1930s, its substantive orientation by Robert K. Merton and Thomas S. Kuhn, and the late recourse to Ludwik Fleck's ideas (e.g. *Denkkollektive*) heralded a *Pragmatic Turn* and many related ones besides. Pragmatism holds that the meaning of a concept lies entirely in its practical consequences and that action is, therefore, the origin of all things. Thinking is an instrument for bringing forth new ideas in the world (Bernstein, 2010). Thus, it signifies an active-creative function of thinking and, in consequence, a rejection of all those concepts in which thinking is seen as a display of transcendent, ontological realities which are merely received in the act of thinking.

This turn is particularly evident in STS, an interdisciplinary, scientific research field established in the 1970s to investigate the interactions between society, politics, and technological development. In the process, well-known scientific approaches such as the Actor-Network-Theory (e.g. Latour, 1993) or the Third Wave of Science (Collins & Evans, 2002) assign a new or higher status to experiential knowledge.

These anecdotes underscore the emergence of a new conception of experience within diverse scientific disciplines, driven by pragmatism and its subsequent resurgence prompted by the mentioned turn. In the following, the concepts of experience of William James and John Dewey, as prominent representatives of pragmatism, will be examined. Their contributions facilitate an active comprehension of experiences imbued with emotion.

3.2.1 William James and the Creative Reality

The notion of experience held a central position in the work of William James (1842–1910). 'It is one of the most important paradigms of modern humanities in the sense of a new critique of consciousness. [...] James [...] formulate[s] [his] critique of consciousness as a radical-empiristic attack on metaphysical concepts of consciousness (Bogusz, 2009: 202, translated from German).'

In his treatise on the fundamental character of the subject-object relationship, James raises the question: 'Does Consciousness Exist?' (1904), initiating a critique of the entrenched dualism between subject and object, a concept long established in philosophy (Russell, 1975). Primarily, he questioned the phenomenon of cognition (Ger. *Erkennen*), which requires a consciousness; everything presupposes the subject as mind or spirit and the object as something materialised. He endeavours to reconcile the divide between rationalism and empiricism through the original tenets of pragmatism, aiming to elucidate experience as a foundational rational element with his concept of 'Radical Empiricism' (1912).

William James, a philosopher, medical scientist, and psychologist, indeed had an interdisciplinary background, and thus, he called for a holistic view that transcended scientific disciplines. His projects were to be interdisciplinary and, indeed, transdisciplinary. James critiqued the conventional subject-object dualism prevalent in prior philosophical thought, arguing that it failed to account for the entirety of life, body, and environment. The division inherent in dualism clashed with James's holistic perspective, which sought a more inclusive understanding of truth. Thus, he contended that only radical empiricism could effectively address this challenge.

The early pragmatists, including Charles Sanders Peirce, William James, George Herbert Mead, and John Dewey, rejected the notion of adhering to theoretical constructs imposed by philosophy, particularly in its moral frameworks, in favour of engaging in experiential inquiry and experimentation to develop theories and acquire knowledge. They advocated for the examination and comparison of *multiple truths* rather than the acceptance of a singular, universally valid truth that imposes a uniform reality on all individuals. This perspective represented a radical departure from the prevailing rationalisation processes of modernity. It can be characterised as a fusion of ontology and empiricism, manifesting as a *Weltanschauung*, or mosaic philosophy (James, 2006: 29), which synthesises diverse experiences that shape our perceptions, realities, and worldviews.

James's perspective on pragmatism evolves into a humanistic framework that prioritises the centrality of human experience. In this framework, reality is not perceived as inherently given but rather emerges from subjective, situational experiences, which may not solely rely on objective facts. However, this does not imply arbitrariness in the world but rather underscores its creative nature, aiming for a pluralistic perspective (see subchapter 3.4.1). This perspective acknowledges that the world is not predetermined but offers opportunities for individuals to exert influence and agency. It posits that everyone is an active participant in a network or mosaic of experiences, thereby sharing equal responsibility for their actions—an idea drawn from James's meliorism influenced by Peirce. For James, this approach represents a means of resonating with the world.

Pragmatism fundamentally posits that the significance of a concept resides in its practical implications. Action emerges as the fundamental driver of all phenomena and serves as both the precursor and objective of all cognitive processes. James expanded upon this notion by formulating a theory of truth that associates the worth of a proposition with its utility. According to James, truth is contingent, context-dependent, and subject to continuous evolution, rendering it only temporarily valid. Consequently, as previously mentioned, there exists not a singular truth claim but rather a multitude of divergent ones—a pluriverse. This conceptualisation underpins the democratic principle, which later served as a method for Dewey to advance communicative, intersubjective experiences.

3.2.2 John Dewey's Experiences as Interactions

John Dewey's (1859–1952) pragmatism, extending from James's principles of resonance with the world and initiation of creative processes, offers a remedial approach and presents an analysis of experience as a wellspring of ideas. Particularly in his later works 'Nature and Experience' (1925) and 'Art as Experience' (1934), Dewey discusses his major philosophical concept. In 'Nature and Experience', often seen as Dewey's metaphysical exploration, the notion of nature is viewed as a preliminary understanding that anticipates the significance attributed to experience. Dewey's conception of nature elucidates that it is not a predetermined entity but rather an ongoing interaction between living organisms, species, and environmental factors within a developmental framework. This evolution is contingent upon openness and situational context, drawing upon its potential from dynamic interactions. A parallel notion emerges in his work 'Art as Experience', which probes into the philosophical underpinnings of art theory, investigating the essence of aesthetics and the interconnectedness of art, society, science, and emotional responses. However, unlike conventional art theory texts, Dewey's approach shifts the focus from the audience to the artist, providing a distinctive perspective centred on the creative act and its transformative potential.

Dewey's perspective extends beyond the artist as a creative individual. He posits that '[t]he intelligent mechanic engaged in his job, interested in doing well and finding satisfaction in his handiwork, caring for his materials and tools with genuine affection, is artistically engaged (Dewey, 1934: 4).' Accordingly, Dewey's concept of activity or creativity is not limited to the realm of art. It can be extended correspondingly and applied and aligned to this work. In this context, creativity becomes a trait of anyone who can consciously engage with their work and environment and who demonstrates commitment and care (Mol, 2008: 50; Tronto, 1993: 102).[4] I will revisit this point later. Ultimately, this premise already hints at Dewey's views on democracy.

Dewey's democratic perspective rejects the differentiation between high and low art. He argues that this separation disconnects society from the comprehensive experience of art and aesthetics. As a result, society loses its intuitive understanding

---

4   Joan Tronto and Berenice Fisher, for example, provide a convincing definition of 'care' that fits and enhances John Dewey's view of the artist's activity and explains how to imagine the engaging work of someone that impacts their surroundings: 'On the most general level, we suggest that caring be viewed as a *species activity that includes everything that we do to maintain, continue, and repair our "world" so that we can live in it as well as possible*. That world includes our bodies, our selves, and our environment, all of which we seek to interweave in a complex, life-sustaining web (Tronto, 1993: 102).'

of daily life experiences in creativity and aesthetics, relegating art to what is collected in museums. However, Dewey primarily refers to the concept of experience in his writings. He differentiates between two types of experiences: one of a somewhat unconscious nature that has no significant influence and another of a unique kind, which assumes an almost esoteric character.

'Experience occurs continuously', John Dewey writes, 'because the interaction of live creature[s] and environing conditions is involved in the very process of living.' When individuals find themselves in conflictual situations, their interaction with the environment brings emotions, ideas, and conscious intentions to the fore, exhibiting different behaviours compared to experiences not consciously perceived. 'Non-consciously' experiences, however, lack clear intentions and structure, starting abruptly and ending prematurely – 'we start and then we stop.' (Dewey, 1934: 35). In both scenarios, the focus lies on the dynamic interplay between an individual and the external world, indicating a state of continuous evolution rather than a static condition. Dewey later elaborated on this concept, introducing distinctions from the aforementioned understanding:

> In contrast with such experience, we have *an* experience when the material experienced runs its course to fulfilment. Then and then only is it integrated within and demarcated in the general stream of experience from other experiences. A piece of work is finished in a way that is satisfactory; a problem receives its solution; a game is played through; a situation, whether that of eating a meal, playing a game of chess, carrying on a conversation, writing a book, or taking part in a political campaign, is so rounded out that its close is a consummation and not a cessation. Such an experience is a whole and carries with it its own individualising quality and self-sufficiency. It is *an* experience (Dewey, 1934: 35 f.).

According to Dewey, there is a distinction between an ordinary and extraordinary experience, with the latter termed *an experience*. Whereby, through distinctive perception, the ordinary can also be 'reactivated' to be consciously perceived. In extraordinary experiences, we encounter the particular, both positively and negatively. We sense a resonance, an activity. Through the new, the unfamiliar, and the non-repetitive, we ultimately arrive at *an experience*. It has a starting point and an end. Therefore, unlike the one before, *an* experience has structure. For Dewey, consciousness is the key to transforming a habit into shock, that is, the impetus, a resistance that becomes usable for changing the existing arrangement of matter (Dewey, 1934: 35 f.).

With us, no two days are ever the same. So, well, of course, some days are more exciting than others. However, especially when we are working towards a milestone, it feels like unforeseen things are happening all the time. That sounds kind of uncontrolled, but it just shows that we are all doing this for the first time.
(Interview from 30/01/2020, Bahar, Physician & Innovator at Health Hub, own translation of the German transcript)

John Dewey interprets this moment of active, conscious experience as the birth of cognition. This cognition catalyses the formation of ideas and their subsequent creation. Dewey views art as the ultimate expression of this process. For him, art is not just a concept but a testament to humanity's ability to enhance life. It continually serves as a fresh foundation for the genesis of new creations. Art bridges the gap between sensory perception, internal needs, and impulses, manifesting them into external forms.

In 'The Quest for Certainty' (1929), John Dewey presents a surprising passage that addresses the formal essence of an object, echoing Karl Marx's perspective while also examining its emotional impact. Dewey introduces a social-emotional dimension, moving beyond mere functionality. He critiques the purely rational functional approach of modern philosophy for its neglect of experience. 'These preconceptions [that materials only follow their molecular properties] are the assumption that knowledge has a uniquely privileged position as a mode of access to reality in comparison with other modes of experience, and that as such it is superior to practical activity' (Dewey, 1929: 103). In his discussion on art, Dewey extends this argument, emphasising the integral role emotions play in connecting us with art:

> Suppose [...] that a finely wrought object, one whose texture and proportions are highly pleasing in perception, has been believed to be a product of some primitive people. Then there is discovered evidence that proves it to be an accidental natural product. As an external thing, it is now precisely what it was before. Yet at once it ceases to be a work of art and becomes a natural "curiosity". It now belongs in a museum of natural history, not in a museum of art. And the extraordinary thing is that the difference that is thus made is not one of just intellectual classification. A difference is made in appreciative perception and in a direct way. The aesthetic experience – in its limited sense – is thus seen to be inherently connected with the experience of making (Dewey, 1934: 50).

The moment of alienation, as previously described, is already inherent. This leads to a reordering of things, prompting a new assessment and reflection on an object's evaluation. These are familiar moments when we find ourselves revising a previous assessment, often accompanied by an astonished 'Oh, I see'. This conscious process questions the preceding one and may lead to a new conclusion, an appreciation, or even a degradation. Interestingly, Karl Marx's description of the commodity fetish (Marx, 1887) and Bruno Latour's 'Factish' (Latour, 2010) both provide similar accounts of this reordering. Ontologically, the focus shifts from the purely physical or (bio-)chemical functioning of an object to what the object evokes in an individual or even a collective. The question arises as to the extent to which it possesses a specific potential for power because the mere 'belief in the belief' prevails and escapes criticism.

Emotions, which Dewey primarily characterises as aesthetic experiences, are integral to his concept of experience. As evident in pragmatism, Dewey discourages a purely modern approach. He views emotions not as fleeting or casual feelings, but as the quality that captures the complexity of experience. According to Dewey, any intense expression that quickly surfaces and fades is a reflex, referred to as an affect by others. This perspective underscores the depth and intricacy of human experience.

> Physical things from far ends of the earth are physically transported and physically caused to act and react upon one another in the construction of a new object. The miracle of mind is that something similar takes place in experience without physical transport and assembling. *Emotion is the moving and cementing force*. It selects what is congruous and dyes what is selected with its color, thereby giving qualitative unity to materials externally disparate and dissimilar. It thus provides unity in and through the varied parts of an experience. When the unity is of the sort already described, the experience has aesthetic character even though it is not, dominantly, an aesthetic experience (Dewey, 1934: 44).

Dewey, as evident in the quote, connects the previously criticised modern philosophy, e.g. the way of thinking adopted by the Cartesian Turn and the previously overlooked aspect of experience. He elevates the status of experience, a recognition it has not previously received in the realm of natural sciences. In this context, he lauds emotion as a 'moving and cementing force [...] giving qualitative unity to materials'. At around the same time, his colleague Emile Durkheim would refer to the occurrence of sensations as 'social or organic solidarity' (Durkheim, 2013), attributing an indicative role to emotion in this context. It determines compatibility and harmony. Through various experiences, emotion bears witness to the degree of qualitative unity. Pertinent to this thought is the fact that from the moment of experience and the ensuing idea in prototype development with its iterative loops, numerous other experiences are gathered from different actors. Emotions evaluate these many experiences and signal the degree to which something should be incorporated as a distinct idea in the innovative artefact and to what extent a 'temporary entity' arises from it (see 3.4 Interim Conclusion: The 'Moral Economy' Around the Artefact).

> Well, I had a similar idea, and there was already something like this here [at Health Hub]. [...] Moreover, Hendrik brought in the patients because we need to know

what they think and whether it works. It is the data from the patients that flow into this, and we evaluate many parameters. [...] The milestone plans are a different story. We have discussions with Tim and the consultants, and there are always ideas that come into it. Tips or something – but mostly, I say that things do not work that way. In the end, it is a collection of many ideas, but what is implemented must ultimately be technically feasible. *(Interview from 04/02/2020, Viktor, Developer at Health Hub)*

3.3 Emotions Constituting the Technological Artefact

This subchapter shows in more detail how emotions can be projected onto things based on one's imagination. The assumption is that it is through emotions that we communicate (Döveling et al., 2010; Gammerl, 2012; Hochschild, 2012). Emotions are part of our actions and language, but they are also formed and shaped by our environment and culture, and vice versa. This subchapter posits a correlation between our imagination, experiences, and emotions, which is evident in our actions. This aligns with the pragmatist approach of *thinking, feeling,* and *acting*.

The primary focus lies on the activating moment triggered by emotion within an individual. The objective aims to demonstrate that *creativity* is awakened in individuals as a result of specific emotional states, constituting an activating moment that stimulates the generation of novel ideas or artefacts. This activation serves as a fundamental prerequisite for individuals aspiring to innovate. Furthermore, the nexus between imagination, emotion, and (re)action crystallises as the relationship with oneself, one's group, and the broader world. Through emotions, individuals assess both themselves and external entities such as things. Emotions possess an inherent value that individuals can transfer, and which can have a different value depending on the 'moral economy' (see subchapter 3.4).

I thought to myself, we have to do something about this. It made me so incredibly angry that this didn't exist yet. It's so obvious and necessary for everyday clinic life. It can't be that there isn't a functioning solution for this. I'm furious. *(Interview from 30/01/2020, Bahar, Physician & Innovator at Health Hub, own translation of the German transcript)*

The ensuing discussion will explore the pivotal role of emotions in the production of knowledge. It may seem counterintuitive to assert that emotions, abstract by nature, influence the construction of knowledge. Yet, the emotional entanglement with and co-creation of facts stand as a central thesis of this discourse. Historically, emotions have been entwined with science, albeit complexly, as they have not traditionally been acknowledged as analytical faculties. Instead, they have been relegated as subjective and irrational, leading to their marginalisation in scholarly discourse (De Sousa, 1987). In the quest for objectivity of facts and science, the concept of the 'Scientific Self' was rendered impersonal, leading to the dismissal of emotions as unscientific (Daston & Galison, 2007: e.g. 199). The disregard for their quality and fruitfulness for science finds a brief pause in the so-called *Emotional Turn* starting in the 1980s (Hitzer & Gammerl, 2013), a movement that persisted for over two decades[5] and has quietly existed ever since. However, it is occasionally taken up anew in the academic debate.

Despite this, a unified canon among the disciplines still needs to be discovered, resulting in ambiguity regarding the precise implications of the Emotional Turn. This imprecision is hardly unexpected, given the historical side-lining of emotions as either analytical tools or research prerequisites—a viewpoint already critiqued by Alexander von Humboldt (Humboldt, 1845: 5 f.). Contemporary studies on emotions almost invariably reference the Emotional Turn, advocating for a renewed focus on their significance.

Throughout the latter half of the 20th century, efforts to incorporate emotions into scientific dialogue led to pivotal shifts in thought. The Linguistic Turn at the 1960s' close marked a departure from sporadic references to emotions, establishing them as a key analytical category. Figures like Richard Rorty and Gustav Bergmann played significant roles during this period. Post-World War II saw the rise of Neo-Pragmatism in the United States, which diverged from traditional philosophical paradigms by focusing on everyday experiences and their linguistic articulations. This movement rejected the pursuit of an ideal language, as previously sought by language philosophers, and instead embraced the empirical study of language's everyday use.[6] It was unique that no ideal language, as demanded previously by the philosophy of language, was to be developed, but rather the empiricist exploration of language and its everyday expressions. This demand reintroduced a novel form of subjectivism. In this regard, the change conferred a new status on language use and elevated thinking and feeling as subjective elements of experience expressed through language, word choice, and prosody. Consequently, pragmatism may be considered a precursor to more recent studies on emotion.

---

5   There are some scholars from psychology (e.g. Klaus R. Scherer, Harald A. Euler, Heinz Mandl) and the neurosciences (e.g. Antonio Damásio, Joseph LeDoux) who date this turn around the 1980s (Hitzer & Gammerl, 2013). However, in the cultural sciences, it is dated around the millennium.

6   During the development of this newly interpreted school of thought, Rorty distanced himself in 1967 in the preface to the anthology 'The Linguistic Turn' from the analytical philosophy and, in particular, from the so-far-established philosophy of language. For Rorty, the turning away from the still-existing epistemology, and thus the attitude that one could interpret from theory to practice, was decisive.

During the 1980s, the field of emotion research gained substantial momentum. In social and human sciences, researchers have recognised a phenomenon known as emotional blindness, both within their subjects and beyond. They perceive this phenomenon as a deficiency that requires rectification. Subsequently, emotion research has significantly risen, solidifying its position in the hard and soft sciences. This progression indicates the importance of understanding and addressing emotional blindness in scientific studies (Schneider, 2016: 7). Psychology, in particular, identified this deficiency early on. During the 1980s, various congress contributions highlighted this unfortunate situation, advocating for a solution. The primary focus was to explore the interdependence of cognition and emotion further. This indicates recognising the intricate relationship between thinking and feeling in psychological studies (Schneider, 2016: 7).

Starting from the late 1980s, researchers began integrating emotion research with evaluation (De Sousa, 1987),positioning emotions not merely as external entities for evaluation but also as foundational elements for decision-making. This integration fostered the development of new rationality arguments, which catalysed a breakthrough in emotion research. The potential lay explicitly in the *rationalisation of the irrational*—a concept long deemed inconceivable—which attracted interest beyond psychology, including anthropologists and other humanities scholars. This interest led to the rediscovery of a field that historical studies have explored since the 2000s (see, for example, work by Max Planck Institute for the History of Science in Berlin from 2008). Consequently, studying the history of emotions experienced a boom, attracting considerable external interest and expanding a previously limited research field. Psychology, as applied science, has since popularised lay literature in guidebooks (Illouz, 2007), and the private self, inner life, and emotions have emerged as new products of capitalism. There is a heightened interest in the exposition of the body, soul, and emotional self, which has also attracted criticism.

Emotions, feelings, or affect serve as testimonies to social structures. This value of the statement wants to be applied here and made fruitful, albeit not to expose the subsequent guarantors, but to understand processes of change.

Therefore, understanding innovation processes and identifying the (emotional) social understanding of innovation in this context is crucial. Knowing the emotional status of a group helps to understand how it is socialised with and thinks about technological developments.

3.3.1 Emotionality with Things

As already highlighted by pragmatism, we experience ourselves as *thinking, acting*, and *feeling* human beings. All three aspects are central to our human existence and social (inter-)actions. The latter, namely feelings, is especially necessary to understand social order and social change (Barbalet, 2005: 178; 2006: 51). This indicates that some feelings only arise through interaction and need provocateurs. Therefore, it is clear that a feeling refers to the nature of a relationship and its structure and imbalances. Feelings convey their *own* 'grammar': they refer to their vocabulary, syntactic forms, and meanings (Oatley, 1993: 341). Emotions give structure and, therefore, construct social practices and convey a way of thinking and feeling in a society or culture. Therefore, this subchapter emphasises the role of emotional construction in developing an artefact, focusing on the central question of how individuals bring it into being.

Figure 2: From Experience to Evaluation

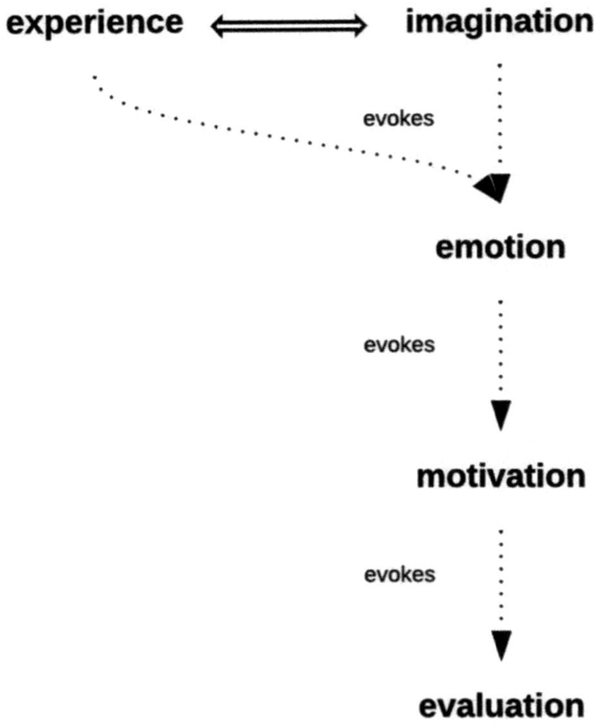

experience ⟺ **imagination**

evokes

emotion

evokes

motivation

evokes

evaluation

Imagination and daily experiences, as previously described and depicted in *Figure 2*, generate ideas that inherently convey emotions due to their unique nature. In simpler terms, our socialisation and everyday lives shape a specific emotional reference system, a product of our reactions to our surroundings. Concurrently, emotions can also catalyse action (Döveling et al., 2010). Emotion, while not directly perceived as an actor, indirectly influences through its motivational force and can later serve as a unifying element at a collective level (see subchapter 3.4). Materiality assumes the role of an emotion carrier, allowing the emotion to gain autonomy.

The idea thus creates an emotionally conceived prototype, a (technological) artefact. In this context, we interpret these artefacts as non-verbal signs that signify their materiality. Emotions emerge from social, material, and, ultimately, sociomaterial relations. Depending on the viewer or actor, interpretations of the idea and the artefact vary, as both provide spaces for projecting the hopes and desires of their creators, viewers, or users. The attributed meaning is situational and depends on the potential representation of the prototype, whether it be a cure, facilitation, or enjoyment. This situational representation ultimately forms the reference system of feelings, similar to what we know from Ludwig Wittgenstein's Private Language Argument (Hintikka & Hintikka, 1986: 244; Wittgenstein, 1977: I, sec. 243).

This reference system of feelings is something learned, formed by the outside and the inside (see subchapter 3.1.1). It is partly established but can evolve, leading to the emergence of a highly differentiated reference system accessed during situations of 'emotional labour' (Hochschild, 2012: 3). What is meant by this is that it is reflective work; one resorts to this system, although this emotional work partly occurs unconsciously because the feelings are not permanently consciously processed but rather express themselves in a way, or we manage them, as Arlie Hochschild indicates:

> Feeling rules are standards used in emotional conversation to determine what is rightly owed and owing in the currency of feeling. Through them, we tell what is "due" in each relation, each role. We pay tribute to each other in the currency of the managing act. In interaction we pay, overpay, underpay, play with paying, acknowledge our dues, pretend to pay or acknowledge what is emotionally due [to] another person (Hochschild, 2012: 4).

It is thus a matter of assessing what we are willing to show or how controlled or uncontrolled we want to be. It is an assessment of the situation and an evaluation of the situation in which we are interacting. However, it is not only a question of what we are willing to give, in the sense of an emotional concession, but also what we consider necessary. It is the exchange of 'goods' whereby the currency, as Arlie Hochschild mentions, is the feelings. The situation is the marketplace of the reference systems developed in each case. Thus, in every situation, we evaluate what something is worth to us and invest our feelings accordingly. This assessment applies to every negotiation situation under the aspect of what outcome the actors hope for. These negotiation processes occur within teams working on a technological artefact. It starts with the idea and the projected expectations of the product's final capabilities. As indicated, this makes the product the result of this emotional negotiation in various situations. Ultimately, it is a process of negotiating what the team around the prototype deems 'valuable': which ideas to implement, what should remain in the final result, or what they can achieve within the given time frame or budget (see subchapter 4.3).

> We always fight [laughs]. If we were a married couple, we'd probably be seeing a therapist. We are in some way because we're constantly being counselled, we're taking courses, and we also have to explain ourselves. Everyone is given space. I think that's a good thing because, in the end, everyone values something different, and we have to somehow work together to get the thing finished and also to make it "sellable". So, emotions are always there, but it doesn't work without them; they are definitely what drives us. If everything was so neutral, ha [laughs], then maybe we would be faster sometimes, but certainly not as creative. *(Interview from 30/01/2020, Bahar, Physician & Innovator at Health Hub, own translation of the German transcript)*

> The "interactional" theorists assume, [...] that culture can impinge on emotion in ways that affect what we point to when we say emotion. [...] I think of emotions as more permeable to cultural influence than organismic theorists have thought, but as more substantial than some interactional theorists have thought [...] [An] emotion is a bodily orientation to an imaginary act [...]. As such, it has a signal function; it warns us of where we stand vis-à-vis [an] outer or inner event [...]. Finally what does and does not stand out as "signal" presupposes certain culturally taken-for-granted ways of seeing and holding expectations about the world – an idea developed [...] on the naming of emotions (Hochschild, 2012: 16).

Several methods can validate this statement, such as examining the diverse terms for 'fear' or 'joy' in various cultures or exploring different narratives within specific disciplines that relate a feeling to their work. For once, it is societies and their cultures that develop a particular language with which we collectively manage our everyday lives. Nevertheless, the many small disciplinary cultures or thought collectives (Ger. *Denkkollektive*), as Ludwik Fleck (Fleck, 1935) calls them, develop their language and, based on this language, bring different expectations to bear on an artefact, in this case. Ludwik Fleck, a microbiologist, medical scientist, and science theorist, posits that a collective forms knowledge and its production. It has no individualistic character, and if so, only in a gathered, collected sense. 'Denkkollektiv' is defined as '[...] as a community of people who exchange ideas or interact in thought, we have in it the bearer of the historical development of a field of thought, of a certain body of knowledge and state of culture, i.e. of a particular style of thinking' (Fleck, 1980: 54–55). How people communicate and their socialisation within their subject or group aligns them in their thought direction or how they perceive something. They adopt a particular perception that makes up their style of thinking. However, with Fleck, there is the distinction between the eso- and exoteric circles in and around a thought collective whereby the esoteric circle consists of experts while so-called 'informed laymen' form the exoteric circle. The informed laymen thus do not form part of the 'inner circle' but rather are an informed group around the thinking collective that knows part of the thinking style or language. This distinction will be important later in the empirical part since these group structures also exist in the context of prototyping labs and makerspaces, namely an inner circle and a group that is indirectly informed or made aware where the latter know part of the content or are informed about the 'spoken language'. It goes even further as these circles are permeable; they can be transgressed and do not have to be self-contained.

Bearing this in mind, how can we exclude the possibility that they do not feel differently based on their expert language and project different attitudes, expectations, and desires onto a product or apply them to it?

> This is a microcosm here. Everyone wants to get something specific—me, a product that corresponds to my idea. Health Hub wants to know that we are developing something worthy of being funded here. The insurance companies want a safe product, which is why we have the Johner Institute on our side [...]. *(Interview from 04/12/2021, Ryan, Physician & Innovator at Health Hub)*

In a moral economy, as we will see with Lorraine Daston, we again encounter Ludwik Fleck's idea of thought collectives. Daston describes the negotiation processes of the esoteric and exoteric groups mentioned above. Furthermore, these negotiation processes and their confounding moments have the potential for emotional conflicts.

To understand how we arrive at value judgements, I will discuss below how emotions help us make judgements in the first place and how the coming together of different emotional cultures in a common place can also determine what we feel.

3.3.2 How Emotions Lead to Judgement

The difficulty with emotions is that they are often devalued as irrational, even though they can influence our decisions daily. In the complex decision-making dynamics, emotions often operate behind the scenes, subtly influencing the process. As a result, attributing decisions to specific emotions can be an intricate task due to their less overt visibility. However, it is essential to take a step back and not expect that a given decision is prompted and determined by a situational emotion. Instead, it is the case that we continuously perceive the world around us through our emotions, and we can thus also make a judgement based on long-term feelings and sometimes on the ideals that arise as a result—furthermore, starting from emotions as feelings are complicated, as I can name a feeling but not necessarily an emotion. Thus, I can feel anxious, sad, or happy, whereby these expressions are feelings. We often assume that emotions are the already processed perceptions of our consciousness, which is why we are all too often familiar with the supposedly conscious handling of naming a feeling. In this respect, it is unsurprising that psychology has been trying to bring order into this emotional chaos for a long time. Three terms frequently cross our path: emotion, affect, and feeling. In addition, there are first and second-order feelings (Archer, 2000: e.g. 197 f.) or 'factive' and 'epistemic' emotions (Gordon, 1987: 45 f. and 65 f.). All these are attempts to distinguish the mere description of a headache from lofty romantic feelings or to have generally explanatory patterns for people's actions. However, as elucidated in subchapter 3.3.1 and reiterated here, emotions, encompassing both affects and feelings, articulate aspects of our attitude, delineating our boundaries, beliefs, values, and objectives (Sartre, 2015; Solomon, 2004). Since Immanuel Kant's time, diverse philosophical and psychological theories have emerged to address individual (value) judgements and decision-making processes. Building upon Franz Brentano's Theory of Judgement, philosophers posit that emotions exhibit *intentionality* (Brentano, 2015; Robinson, 2005). As described in subchapter 3.1, emotions are directed at something, and therefore – according to philosophy – so is judgement (Gordon, 1981; Lazarus, 1991; Nussbaum, 2003; Solomon, 1993). Love and hate, as exemplary emotions, necessitate an object for their direction; such feelings cannot exist without a focal point. Moreover, judgement itself requires a subject of evaluation; there is always something upon which I pass judgement.

The perspective proposing that emotions constitute evaluative judgements may appear extreme, given the necessity of other components, such as action tendencies and physiological changes for the experience of emotion. Nonetheless, emotions must inherently encompass some form of judgement (Robinson, 2005: 11).

Whether phenomenologists, existentialists, or psychologists, all are certain of at least one thing: judgements are always situational statements. If they involve emotions, they are evaluative judgements that indirectly name desires, values, interests, and goals. Judgements are situational because individuals consistently evaluate situations in different ways. However, evaluations occur in a moment of interaction with my surroundings, representing a convergence of inner and outer dynamics. Thoughts, associations, and memories intersect with the external world, prompting an immediate confrontation. The perspective that emotions constitute evaluative judgements might seem extreme, given the additional components required for the experience of emotion, such as action tendencies and physiological changes. Nevertheless, emotions must inherently involve some form of judgement.

'Richard Lazarus, for example, has claimed that the relevant "judgement" that forms the "core" of an emotion is always an "[a]ppraisal of the significance of the person-environment relationship"' (Robinson, 2005: 12–13). However, there is no consensus concerning the relationship between emotions and judgements. 'Some think that emotions are identical to judgements, others that judgements are sufficient for emotions, and others again that judgements are a necessary condition for emotions but not sufficient' (Robinson, 2005: 14). Still, regarding emotions, as judgements seem misguided, simply transferring emotion directly into a judgement proves inadequate. Individuals can judge situations independently of their feelings by consciously distancing themselves from them. Subscribing to an 'objective' rationale, contradictory to personal feelings, is also possible. However, a direct relationship between emotion and evaluation persists, as consciously distancing oneself from emotions requires effort, albeit inherently linked to the self, albeit inverted. Such efforts entail reviewing one's emotions to facilitate evaluation. Without this, one may assume a linear relationship between emotions and resulting judgements.

Furthermore, it is crucial to recognise that emotions inherently convey a value-based understanding, shaping everyday decisions. However, as previously outlined, this does not automatically justify every action; instead, it involves individually considering the environment and one's perception of it, as elucidated by Arlie Hochschild.

3.4 Interim Conclusion: The 'Moral Economy' Around the Artefact

In 1995, Lorraine Daston, a historian of science, proposed a solution to the paradox of science with her concept of 'moral economies' (Daston, 1995). This paradox relies significantly on the idea 'that science depends in essential ways upon particular constellations of emotions and values (Daston & Galison, 2007: 3)', although there exists an expectation for science to embody rationality and objectivity. This entanglement is what she later picks up again together with Peter Galison:

> All epistemology begins in fear – fear that the world is too labyrinthine to be threaded by reason; fear that the senses are too feeble and the intellect too frail; fear that memory fades, even between adjacent steps of a mathematical demonstration; fear that authority and convention blind; fear that God may keep secrets or demons deceive (Daston & Galison, 2007: 372).

It thus becomes apparent that neither sciences (nor scientists) are free from subjectivity and, therefore, neither from emotions nor emotional attachment. In this regard, it exposes evidence as a 'child of our time'. The categories in which we think, the schools and styles of thought we develop, and maybe even the evidence we create (see also Latour's concept of *Factish*) are born out of time and the way of knowledge production of a specific group (as in Fleck's *Denkkollektiv*). Through empirical investigation, the concept of objectivity appears elusive. Yet, in a positive light, there exists a set of objective principles that have been developed and persistently applied, albeit subject to potential debate.

Nevertheless, the name Daston chose, namely the *moral economy*, might need some clarification. To understand what a moral economy means, we must refrain from using the term economy and its understanding in the economies themselves. Instead, we have to go back to the word's original sense, which refers to a forum where people exchange 'moralities' or rather 'moral values', which generally refers to our expectations. Precisely, it 'is a web of affect-saturated values that stand and function in well-defined relationship to one another' (Daston, 1995: 4); an assemblage of diverging affects and emotions (hopes and fears) that animate their development and politics. *Moral* in this regard refers to both affective and normative dimensions, and *economy* is used in its expanded sense as a 'balanced system of emotional forces' (Daston, 1995: 4), that is, the system of emotional forces and affect-saturated values that surround and constitute a 'thing'.

This definition is a contingent and, at the same time, dynamic system, not a means to an end, and yet, the structure and the modes of action of a moral economy underlie a certain logic. 'Not all conceivable combinations of affects and values are, in fact, possible. Much of the stability and integrity of a moral economy derives from its ties to activities [...] which anchor and entrench but do not determine it (Daston, 1995: 4).' This is when actors of knowledge production become relevant, whereby it does not yet indicate anything about their motivation to 'do' science. Moral economies are a relevant and indispensable part of science. They significantly influence scientists in terms of how and what to think and what topics to work with, including their decision-making and the objects they examine. Moral economies tell us how and why scientists pick or choose particular objects, which explanations they trust, and which habits and methods they use or develop. In the history of science, several examples demonstrate how much disciplines depend on the above-mentioned aspects. At this point, one can refer to the so-called 'Science Wars', which repeatedly serve as an example to illustrate disciplinary differences within the sciences and argue about schools of thought, methods, and language. Frequently cited examples are the positivism dispute between Karl Popper and Theodor W. Adorno or the Sokal affair, or rather crisis, which was caused by and named after Alan Sokal. The latter, in particular, triggered a debate on intellectual standards for the social sciences, specifically devoted to the 'Science Wars'. As it turned out, this was a fundamental clue as it could not have demonstrated better how great the differences between the hard and soft sciences were.

By examining the moral economy of an artefact, questions about the thought patterns of stakeholders and actors can be answered, including: How are relevant actors emotionally shaped? What do they see as relevant? When do they judge something to be relevant to research, or how and when do they promote an idea?

Before initiating an initial analysis, it is imperative to pose two key questions. Firstly, how do the actors locate one another and mould their negotiation procedures assuming a shared interest in accommodating their respective ideas? Secondly, after reaching an agreement and commencing collaborative efforts, how does an artefact emerge as the cohesive force among these transdisciplinary actors? In other words, how does the prototype, akin to Durkheim's notion of a social glue, facilitate the navigation of milestones, overcome obstacles, and mitigate discord within the team?

3.4.1 The One and the Many Ideas

Before empirically addressing the aforementioned questions, I aim to contextualise them within a theoretical framework that initially engages with a philosophical problem—a quandary that arguably stands as one of the most substantial of all time and continues to hold prominence in philosophical discourse (Savransky, 2021a: 4). The problem of 'the one and the many' discusses the question concerning to what extent we speak of a reality, whichever and whose it may be, i.e. that of the 'West or the "rest"' (Savransky, 2021a: 4) as a unity or a plurality and how the two relate to each other—or not. We cannot rule out that '[…] we base our conventional notions of what is real on a belief that we interact with the world as individuals separate from that world (Escobar, 2020: 2).' We typically examine our world ontologically, relying on our received teachings and formulating our ideas and conceptions accordingly. Even on a small scale, I would like to illustrate how the problem of 'one and the many' can become visible in an idea as an act of individual access to a world and why it remains a problem.

It is precisely in developing ideas and the subsequent development of prototypes that it becomes evident to what extent *one* idea may become *many* or vice versa. However, in the end, a given narrative tells us about a unified one, in which one cannot necessarily divide the plural aspects inherent to it. In this subchapter, two starting points have one thing in common: a problem—a problem unifying both points.

On the one hand, this refers to a problem that we detect in everyday life, through – as described earlier – our experience as William James, in a little anecdote in his talk from 1907, 'The One and the Many', indicates:

> I have sometimes thought of the phenomenon called "total reflexion" in optics as a good symbol of the relation between abstract ideas and concrete realities, as pragmatism conceives it. Hold a tumbler of water a little above your eyes and look up through the water at its surface – or better still look similarly through the flat wall of an aquarium. You will then see an extraordinarily brilliant[ly] reflected image say of a candle-flame, or any other clear object, situated on the opposite side of the vessel. No candle-ray, under these circumstances gets beyond the water's surface: every ray is totally reflected back into the depths again. Now let the water represent the world of sensible facts, and let the air above it represent the world of abstract ideas. Both worlds are real, of course, and interact; but they interact only at their boundary, and the locus of everything that lives, and happens to us, so far as full experience goes, is the water. We are like fishes swimming in the sea of sense, bounded above by the superior element, but unable to breathe it pure or penetrate it. We get our oxygen from it, however, we

touch it incessantly, now in this part, now in that, and every time we touch it we are reflected back into the water with our course re-determined and re-energised. The abstract ideas of which the air consists, [are] indispensable for life, but irrespirable by themselves, as it were, and only active in their re-directing function. All similes are halting but this one rather takes my fancy. It shows how something, not sufficient for life in itself, may nevertheless be an effective determinant of life elsewhere (James, 1922: 127 f.).

This quote illustrates William James's abstract way of explaining that we are, after all, limited by our perspectives. Although new experiences constantly enrich these, we are ultimately unable to see a 'problem' other than the one we recognise, with which we are already familiar. For this, we need others to replenish our worldview or access the world through their world.

However, if individuals did not encounter problems in their daily lives, the need for a unification process would not arise. I propose that individuals identify various problems they intend to solve through their own ideas or the ideas on which they collaborate. Therefore, as previously outlined, they anticipate the idea, or ultimately the product, to serve as a solution for the issues they perceive. Therefore, instead, the second problem or extended problem unifies the many ideas, and a development process to find consensus is actually possible.

Thus, as discussed earlier, for the negotiation processes of prototypes, we need to start from a forum where idea contributors, incubators, developers, financiers, consultants, and sometimes insurers –when it comes to medical prototypes – meet. In the case of the latter, med-tech certifiers are also necessary as they are familiar with the regulations that must be adhered to so that a product can eventually reach the market.

In brief, we start from a forum where actors with different professional backgrounds meet and cannot constantly assess each other's work content. For example, some are medical doctors; others are economists, consultants, and developers, to name but a few. They all have different professional knowledge, interests, and goals. Ergo, the developer tends to focus on feasibility in the context of product development. Contrastingly, the medical scientist begins with a concrete, idealistic idea, as the empirical section of this study will elaborate upon in greater detail.

If this serves as the starting point, it also implies that all initially involved parties must first brief each other and are obligated to exchange information about their respective subjects, knowledge, and expectations to initiate collaboration. Theoretically, this constitutes the inaugural 'iteration loop', a term typically reserved for prototypes as they progress through various developmental stages (see Chapter IV). However, we can commence from these developmental phases even before creating the first model, given that we are in a social setting where initial assessments are made through emotional engagement, as previously mentioned. People encounter one another, react, and evaluate one another.

In his fourth lecture, 'A New Name for Some Old Ways of Thinking', at Columbia University in January 1907 (James, 1922: 127 f.), William James postulated variety instead of one unity to be able to understand the world. He presupposes human curiosity as their will and their power to understand the world and seeks compromise, as before, between realists and empiricists. Hence, we must regard the process of materialising the idea as a multiplicity or pluriverse, culminating in creating a tangible artefact. It involves 'cultivating "a world of many worlds"' (de la Cadena & Blaser, 2018; Savransky, 2021b). To create a cosmos where all experiences converge, forming a realm of experience. It thus remains an 'ongoing and unfinished' (Savransky, 2021b: 143) process.

The assumption presumes that we are dealing with 'problematic' constellations with positions that show discrepancies among themselves. An 'aching gap' (James, 1902: 259), as James calls it, has to be filled, and through the constant effort and endeavour with the affected ones, they resonate, and a 'tensional activity' (Savransky, 2021b: 146) occurs. In this respect, pluralism, however small a role it may seem to play in everyday life, becomes a social risk we cannot avoid taking.

3.4.2 Esprit de Corps in the Moral Economy

Once this space, meaning the pluriverse of many experiences has been created so that there is increasing unity about what the artefact can, should, or must do, the artefact opens up new possibilities. The moral economy, as a place of negotiation, as a group that informs each other and opens up a world of thought and feeling, now finds a new identification in the artefact. The moral economy shows *esprit de corps*. Nevertheless, of course, time is required, and, apart from this, it is also essential to ask how a feeling of solidarity or loyalty comes about and what the nature of this feeling is. Is it because of pure goodwill, sympathy, pressure, or mutual dependence that the group works together long-term, and how stable are these structures?

To examine this, I first refer to parts of Emile Durkheim's social theoretical concept to describe the group dynamics at this moment. First of all, it must be assumed that an initial source of inspiration or an inventor comes from an 'archaic group' (Durkheim, 2013). In the broadest sense, this means that the source of inspiration or person comes from a familiar structure ('archaic group'), which is particularly evident in terms of its clearly defined values, norms, and evaluation patterns. According to Durkheim, the group is also characterised by its members' tight integration, who have little contact with other groups or, in this case, disciplines. It serves as a strut, although it now only initially unites similarities. As Emile Durkheim calls it, this so-called 'mechanical solidarity' (Durkheim, 2013: 57–58) must overcome itself to transform into a moral economy as a field of negotiation and the possibility for new group affiliation. Group members come from their original area to a new setting, e.g. the incubator. Before they experience and reproduce their values through the original small group affiliation, they are firmly integrated into their system and their group and are rarely in touch with other professional groups. Their original collective consciousness, as Durkheim calls it, describes their belonging through language, morals, customs, etc., in brief, their culture, which as an identity-forming element holds the group together and shapes it. It is handed down and passed on. Collective consciousness creates a framework of mutual feeling and evaluation, i.e. a joint attitude. Durkheim describes the rejection of any threat to the system. If such a dissolution takes place voluntarily and constellations are created anew, the possibility for a moral economy can emerge.

> If you think about it, how many medical students have you met during your studies? Not so many. Most of them have their own campus, usually located in the university hospital. That's usually on the other side of town. And, of course, you also have a circle of friends that is so exclusive. I didn't know any software developers or technical designers when I came out of my studies. I didn't have these people in my circle. I just had other doctors. But you can't found a start-up with five doctors, not for MedTech. And then we started here with external contractors. But that wasn't so ideal. They only want the money, and what they deliver is always the minimum version. And then, by chance, we got Viktor, our technical developer. He studied computer science at MIT and has five years of experience in designing wearables. And then we got someone for the business administration part. My sister is already back in her studies; she dropped out again [of the team]. *(Interview from 30/01/2020, Bahar, Physician & Innovator at Health Hub, own translation of the German transcript)*

Thus, to enter or survive in this new group, a much more intensive form of communication is necessary. The former firm integration of the archaic group no longer exists in the new group; Durkheim calls it 'modern society'. The values that were once clearly outlined are no longer such, and the new members, who come from many fields and disciplines, can determine a new consensus of values. However, in the beginning, the original codes of a group are questioned in the new structure. In this respect, the 'collective consciousness' is no longer a traditional one. We do not find any established structures yet many 'culture clashes' that cause mutual irritation. The new system is comparatively much more fragile, and collective consciousness is not present initially. The possibility of disengagement arises, i.e. the feeling of not feeling bound to the group in combination with wanting to find one's way alone and the urge for individual spaces for action is evident. The potential lack of attachment can become a problem if no 'solidarity' develops, although 'loyalty' might be the more suitable term in this context. The solution to address this challenge is to provide clearly defined tasks for each individual and involves developing harmonious interaction between the individual team members. Under these conditions, a new 'moral' emerges or a feeling of belonging in the new group. The crisis of the group, or the problem of lack of commitment described above, can be intensified by many different expectations, among other things. The desires and expectations, which are linked to an explicit value system, must be communicated accordingly so that they can finally be united.

The group builds a social bond through the artefact and because of it. Although they still need to be perceived as individuals, as they maintain their background of origin, they might develop an inspiring enthusiasm for each other's work. While this does not mean that there are no longer disputes, quarrels, and anger among the actors, in the best scenario, the actors develop a shared narrative which indicates that they are developing common 'social facts', ergo socially-determined behaviour (Durkheim, 2013: 270) for their group, meaning common morals, values, and norms. While they do not initially share these social facts in the group, this grows over time as they work together. A mutual dependence develops within the group, from which, in the best case, trust can grow, as Emile Durkheim describes it. This esprit de corps can also be distinguished from a collective consciousness or mechanical solidarity since they do not share the same experiences at the beginning. Thus, my starting point is that of individuals who develop and feel a form of group belonging and reliability towards each other. This expresses itself in the ordinary everyday working life, in conversations with and about each other and ultimately, in a narrative and founding myth that is shaped and formulated over time or is told at the end, based on success.

> Relationship of trust. That's great and very positive. And I also find it very emotional. Positively emotional. So, it's simply fun. And that is profitable for both sides. And then it also starts to become a togetherness. It's also a relationship that you enter into over time. I, at least, enter into a relationship for a time. But it can also be negative! That's always when, yes, I would put it down to trust. If the people we are looking after, yes, I called it resistant to advice earlier. This is often coupled with arrogance. With an inability to put one's own personality behind. That can tip over into arrogance. *(Interview from 13/07/2020, Felix, Consultant at Health Hub, own translation of the German transcript)*

An empirical question will be to what extent 'there are very general and indeterminate ways of thought and sentiment, which leaves room open for a growing variety and of individual differences' (Durkheim, 1972: 145). Beyond this, it will be examined whether trust – which is highly relevant in both the entrepreneurial literature and empirical studies – entails the reliability that one may initially assume.

VI. Innovation-Making
The Construction of Value

> In August 2021, a significant scandal disrupted the Californian start-up landscape. Renowned media outlets across the United States, the United Kingdom, and Germany published headlines such as 'Rise and Fall of Theranos Founder Now on Trial' (Business Insider, 31 August 2021), 'Selling a Promise: what Silicon Valley learned from the fall of Theranos' (The Guardian, 30 August 2021), and 'Fallen Founder' (Die Zeit, 31 August 2021), raising public interest in the event. The centre of this upheaval was Theranos, a medical technology company once considered a pioneering entity within the illustrious Silicon Valley. Founded in 2003 by Elizabeth Holmes, who was likened to Steve Jobs and lauded as a young, self-made female billionaire, Theranos reached a valuation of one billion US dollars. Its signature invention, the 'Edison' machine, promised to revolutionise medical diagnostics by detecting multiple diseases from a single drop of blood—an unattainable advancement in medicine.
>
> However, the lustre of Theranos began to tarnish as it encountered sustained challenges to its credibility leading up to the scandal. 2015, the Journal of the American Medical Association (JAMA) admonished Theranos for failing to disseminate research findings through peer-reviewed journals. Subsequently, The Wall Street Journal published a series of exposés alleging that Theranos relied on commercially available technology from other large companies rather than its own for blood testing. This revelation sparked further scrutiny, and in 2016, the Centers for Medicare & Medicaid Services (CMS) rescinded the operational and licensure privileges of Theranos' blood testing laboratory in California for two years. This action prompted the United States Securities and Exchange Commission and the California Attorney General's Office to investigate the company's practices.

Although this report alone cannot represent innovation development or a Western innovation culture, it has a particular advantage over a real success story because the series, podcasts, and coverage of this unprecedented swindle deconstructs the founding myth of a company. Such stories are usually 'superior' as how innovation is written about or presented leads to the impression that technical development has followed a straight, rational path from the past to the present (Bauer, 2017: 9). Deconstructing innovation myths is generally challenging. This difficulty intensifies in cases involving successful end products, frequently portrayed as having a straightforward development process. However, this is rarely the case. Technology development is a complex process subject to many circumstances and has a great deal of compromise character.

Theranos is one famous example of failure, which will be a visual example in the following subchapter. Firstly, however, it is necessary to look at the various structures of innovation as spaces for possibilities and how those possibilities are narrated and perceived. Therefore, Chapter IV discusses the environments, proclaimed culture concepts, and opacity of innovation. Further, it goes into detail in examining innovation narratives and performances. Third, it examines the construction of values and the influence of evaluation practices on innovation.

4.1 Structures of Innovation I: Spaces for Possibilities

Introductory Vocabulary on Innovation and its Definitions

Before proceeding with the subject, this section will outline the vocabulary commonly associated with innovation culture and the start-up scene. I will interpret these terms as follows:
Business Angel = Private investors who invest their own money, time, or professional skills and networks in young companies (start-ups) in the hope of financial gain and thus participate in the risks and opportunities of the company's development.
Incubator = The term incubator originated in medicine, referring to preterm infants' receptacles. In the entrepreneurial milieu, it refers to the services and institutions available to entrepreneurs while establishing their businesses. Such incubator facilities are frequently public entities, such as technology centres, with ties to venture capital firms or business angels. Through an incubator, founders can access expert counsel, training, or coaching; they typically receive assistance with the required infrastructure, including office space, hardware and software. Furthermore, the incubator helps founders to access a supportive network. In return, the incubator gets shares in the company if it succeeds.
Iteration = A procedure for the step-by-step solution of a given problem, meaning a non-finished prototype. With the help of a first approximate solution, i.e. further development, additional approximate solutions are developed. These sequences or approximations are called loops and resolve the problem through succession.
Makerspace; Innovation lab; Prototyping Space/Lab = Institutions that rent out workplaces to teams who want to realise an idea. These spaces are usually the first port of call for entrepreneurs, founders, or inventors if they still need to be engaged with an incubator. In ad-

> dition to the workstations, these 'labs' often also provide equipment such as laser cutters, 3D printers, workbenches, and associated workshops, as well as other areas that enable the first distribution of products. These prototyping or innovation labs are often networking places for young founders who can exchange ideas.
>
> Living Lab (sometimes also called real-world laboratory, depending on the goal and definition) = These 'labs' are more extensive than an innovation lab and often refer to a whole region, city, or quarter. Usually, the region holds several projects with a particular mutual goal, e.g. sustainable mobility. Living Labs can function as industrial areas that host testbeds for schemes prior to their broader public testing.
>
> Milestone Plan = A plan that regularly checks the completion of an activity at a specific point in time or event within a project's framework.
>
> All extensive projects, especially those involving software, start-ups, or incubators, generally use milestones to ensure they meet the schedule, cost, and quality requirements specified in the project plan.
>
> Prototype = From the Greek 'protos' meaning 'first' and 'typos' meaning 'archetype, model, form', a prototype is a model designed in science and business to show the essential elements or functions of an imagined and desired component or product. Users employ prototypes to check ideas, test reactions, and find sponsors, aiming to demonstrate the principal feasibility of a concept. Prototypes hold a significant role in technology and computer science.

In 1972, the *Club of Rome* published its State of the World report entitled 'The Limits to Growth. A Report for the Club of Rome's Project on the Predicament of Mankind' (Meadows et al., 1972). Despite already being somewhat dated, the report points to current problems. It serves as the counter-project to the economic liberal attitude, which states that all that is needed is the right innovations to continue to create constant economic growth. Sometimes, these are the only existing opinions on economic growth: its limits, success, and failure. There exists no intermediary stance. Notwithstanding confirming the Club of Rome's view in recent studies and data, buzzword innovation holds its own. This term benefits from a positive, forward-looking narrative and serves as the keyword for economic and infrastructure success with assumptions that their development functions linearly (Reinhart, 2016).

In addition, magazines and newspapers have promoted the 'innovation' section for years. Business and investment websites that only deal with the latest technologies to know when it is the right time to invest 'in the future' are no longer only for brokers but are made available to the general public. Additionally, not only the economic liberal parties are open to promoting investment in new technologies since, due to new societal challenges, the focus on innovation is also shifting. Hence, investment in the same becomes mandatory.

Despite the favourable connotations associated with the term 'innovation', replete with its narratives and myths, and its role as a facilitator for securing both private and public funding, the definition of innovation remains nebulous. In literature and official pronouncements, finding a concrete definition that everyone can relate to is difficult. Ultimately, it is precisely the diffuse and confounding nature of the term that makes it so attractive. For some, it provides access to funding and networks. However, it is impenetrable and difficult to grasp for those unfamiliar with the term. All experience the same ultimate difficulty and fascination: it carries many expectations, lacks uniform description, and ultimately necessitates fulfilling promises. Despite all the ambiguity, one understanding remains untouched, namely that innovation is new or at least novel. At best, it is disruptive and does not involve imitation (Godin, 2017), although the latter aspect is not unconditionally necessary. Here, too, the definitions are not unanimous. The best-known descriptions of what is 'innovative' are those of the Austrian economist Joseph Schumpeter (1883–1950), whose work has been experiencing a renaissance since the 1990s. He defines 'new combinations' as a different way of conveying a product as innovative (Schumpeter et al., 2006).

Even though Schumpeter's definition has influenced many others, I draw on a different clarification, which is helpful precisely because of its precision and brevity; it keeps diffusion in check, and, in this conciseness, it does justice to the term to facilitate working with it. Hence, I refer to a German perspective by Reinhold Bauer from his book 'Gescheiterte Innovation' (Failed Innovation), which reads as follows:

> An innovation is [...] the first economic exploitation of a new problem solution. Essentially, it is irrelevant what kind of solution it is: It can be an organisational change, for example, within a company (organisational innovation), a change in the way a product is produced (process innovation) and/or a change in the manufactured product itself or the completely new introduction of a product (product innovation). [...] The product or process does not have to be new in a fundamentally global sense ("objective innovation"); it is sufficient if the exploitation is a first for the innovating subject or subjects ("subjective innovation") (Bauer, 2017: 11 f., my translation from German).

Bauer presents the three common types of innovation in this definition: 'organisational innovation', 'process organisation', and 'product innovation', describing the standard categories into which innovations are usually classified and thus serve to explain their character.

Ultimately, the concluding remark merits attention, articulating that 'subjective innovation' suffices for an entity to be deemed innovative. Thus, the primary determinant of innovation is the perception of novelty, irrespective of its actual originality. This point may allude to the need for a universally accepted definition of innovation. Moreover, it underscores the notion that innovation does not inherently entail novelty; instead, it represents the timely integration of an idea or product into a society prepared to embrace it (e.g. Akrich, 1992; Bijker et al., 2012; Hoffman & Marz, 1996; Urry, 2016). This observation builds on the close connection between society and the economy, especially concerning innovation (Reinhart, 2012).

Even if structures, processes, and products can be categorised, this only says something about their development, success, aberrations, hurdles, and frequent failures unless their daily conditions are exposed. Subjective innovation is ultimately the term that opens the door to further examination at this point because, as mentioned, it refers to the nature of compromise in a certain way. Since products and processes are guided precisely by these compromises and circumstances in economic, social, political, and cultural terms, it is a permanent negotiation process among different actors (see subchapter 3.4), taking place in iterative loops in designated places. Therefore, the innovation structure leads to the abovementioned circumstances and compromises by actors within the same categories that evaluate existing ideas that might lead to innovation. The structures are manifold. I will, therefore, begin by describing the spatial configurations, which, through the mode, are arranged to guide processes in specific directions. This description already tells us how innovation is currently understood and what sociopolitical interest forms the basis of the type of innovation. This understanding gives rise to a culture that becomes emotionalised and functionalised. Looking at the first external conditions, such as spatial structures, the emergence of culture and the opacity that comes with it, provides a preliminary understanding of how innovation as a mode of practice constitutes a mode of feeling.

4.1.1 Creating a Creative Environment

In contemporary discourse, a tendency exists to overlook societies characterised by innovation. They epitomise evolutionary development at the societal and economic levels. In the last few years, one could increasingly observe how 'Creative City Quarters' (Florida, 2004), 'Living Labs' (Ballon et al., 2015; Bulkeley et al., 2018; Picard, 2017) and 'Sustainable Futures Initiatives' (Dixon et al., 2018; Frantzeskaki et al., 2018) in regions and cities were developing. The number and diversity of these creative places are evident. Is this because of the once gloomy picture painted of Europe's innovative capacity in the early 2000s (see: Nowotny, 2010) and/or the entrepreneurial understanding of science, which has entered into a pact with the economy as an innovation driver (Reinhart, 2012)?

Regarding the myriad structural changes, Europe exhibited a significant demand to enhance its economic spheres, manifesting across various contexts and entailing a broad imperative for creativity. Primarily, one must always consider the relevance of socially significant discussions. These topics often refer to digitalisation, demographic changes, climate, or war. All of them reorder multi-lateral relations and provoke new conducts and demands. The revelation of innovation *dispositifs* demonstrates that creativity and its execution space are boundless.

Furthermore, various forms of innovation also accompany the places of creativity, and the term 'innovation' has been universally discussed from past years to the present (e.g. Färber et al., 2008). Hardly any country, company, or university can do without it. The association of attributes such as *openness, social relevance*, and *sustainability* has emerged in recent years. The economic concept of innovation is gradually dissolving, and for over two decades, a much broader, more open vision of innovation has emerged (e.g. Meissner et al., 2017). These open visions remain key interests, not in the sense of less production-increasing measures or measurable success on the markets, but in the sense that creativity finds its space in various sectors that have not intentionally established a connection to market interests, e.g. makerspaces.

These conditions are no coincidence but are demanded and desired through public calls from the European Commission (EC) (see the *Innovation Union* initiative from 2010 onwards), regions, or cities. In recent years, the imperatives of innovation have gained significant prominence, compelling not only governments to respond in exceptionally modern and innovative manners (Farias & Wilkie, 2016; Hutter, 2016; Pfotenhauer, 2017) but also challenging and necessitating societies at large to do the same:

> The Innovation Union will focus Europe's efforts – and co-operation with third countries – on challenges like climate change, energy and food security, health and an ageing population. It will use public sector intervention to stimulate the private sector and to remove bottlenecks which stop ideas [from] reaching the market (Press Release by European European Commission, 2010).

With this quote, the view on so-called *Grand Challenges*, such as climate or demographic change, is shifting. The transformation in focus mentioned in this quote and the resulting values and practices, i.e. adaptions in behaviour, can be understood as a cultural change. Therefore, different forms of dealing with innovation, specifically innovation cultures – whether nationally generated or institutionally cultivated, as in companies – can be identified. From this, it becomes apparent that it is not reasonable to disconnect macro-, meso-, and micro-levels on an analytical level as they are intertwined. Therefore, they should be referred to as one another (as seen, for example, in the study by Akrich, 1992).

In this context, it naturally follows that the new steering mechanisms, both within the scope of this work's empirical research and more broadly, are designed to nurture particular forms of creativity and set up an initial framework. These mechanisms should establish closer links between the economy and current social debates to test solutions in real-world environments within an experimental framework.

First, regions often fund politically intended *Living Labs* with the hope of uncovering sustainable solutions to region-specific problems through unconventional methods (Bulkeley et al., 2018; Keyson et al., 2016; Wissenschaftlicher Dienst des Deutschen Bundestages, 2018). This means focusing on transfer processes, which in these cases are often identified as so-called 'citizen science' (Irwin, 1995). These citizen scientists use their generally diverse lay knowledge to participate in research projects through observations, raising questions, or active engagement in data analysis (e.g., Bächtiger et al., 2018; Bryan & Tobin, 2019). Ultimately, the closer connection should not only verify societal and/or economic needs, but at the same time, it should also ensure societal and economic success. For example, the former German Ministry for Economic Affairs and Energy (Bundesministerium für Energie und Wirtschaft; BMWi) argues that:

> The coalition agreement sets out the goal of promoting living labs [Ger. *Reallabore*] and experimental spaces in a wide variety of thematic areas. Against this background, the German Federal Ministry for Economic Affairs and Energy intends to strengthen living labs as a cross-cutting instrument of innovation policy. In December 2018, a comprehensive living labs strategy was presented for this purpose, which is based on three pillars (Bundesministerium für Wirtschaft und Energie, 2018: 14, my translation from German).

The three mentioned pillars refer to 'innovation-open regulation', 'networking and information', and 'initiating and accompanying living labs' (Bundesministerium für Wirtschaft und Energie, 2018: 14). As the first pillar suggests, this is about creating flexible innovation spaces that are not subject to legal regulation and yet find a legally secure framework. The second pillar builds on business, science, and administration networking, whereby the main focus is again on legal safeguarding liability and competition issues. The third pillar ultimately refers to anchoring practice and implementing possible innovations outside their testbeds.

Further, these new structures are also echoed at universities, colleges, privately in cities, or by profitable companies that can afford to set up a 'playground' they call a *makerspace* or *innovation lab* (e.g. Davies, 2017). These can be described as shown in the following exemplary extracts (selection):

At the Technical University Berlin:

> The DAI-Labor and chair in "Agententechnologien in betrieblichen Anwendungen und der Telekommunikation" [Agent technologies in operational applications and telecommunications] managed by Prof. Dr. Sahin Albayrak at the Technical

University of Berlin, explores and develops technologies realizing a new generation of systems and solutions – "Smart Services and Smart Systems". The DAI-Labor's goal is to test its custom solutions in a real-world environment and get users in contact with it (DAI-Labor, 2021).

At a private makerspace:

With over 2,000 members in Berlin, Salzburg, and Vienna, we are the largest maker community in Europe. For you, this means concentrated know-how and the opportunity to exchange ideas with makers from a wide range of disciplines. In our regular tours, training sessions and workshops, we pass on our know-how about digital production! (Happy Lab N.A., 2021b, my translation from German).

At a company's makerspace:

The Bosch IoT [Internet of Things] Campus is one of our locations worldwide. More than 300 associates work at the campus in Berlin-Tempelhof – mainly on projects related to the Internet of Things and digital transformation. Our experts advise and support customers in the development and implementation of projects for connected solutions.
The Bosch IoT Campus is more than just a normal office: it brings together the entire IoT ecosystem in one place. The strong team spirit contributes to the unique atmosphere of the campus. In addition to external customers and partners who use the campus to work on projects, various Bosch divisions are also based here. You can also book many of our premises for your events (Bosch IoT N.A., 2021a).

In all these spaces, the provision of infrastructure minimally imposes rules while simultaneously striving to cultivate an optimal environment for the emergence of innovation. In the spirit of 'Constructive Technology Assessment' (CTA) (Rip et al., 1995), the experimental space should be isolated but implemented in networks to establish an exchange with the outside world. It represents the fine line between privacy and the necessary disclosure in favour of innovation and its possible application areas. Despite advocating for openness, labs and spaces must safeguard their privacy. Given the fragile nature of innovation, it necessitates protected environments: like everything that develops, it finds itself in uncertain spheres.

4.1.2 'Culture(s)' in Innovation-Making

These protected spaces, be they named start-ups, incubators, or makerspaces, often talk of 'culture', meaning 'their' culture and corporate culture. The inflationary use and occasional misuse of this term, along with the emergence of hyphenated cultures and neologisms, merely contribute to another aspect of a broadly defined cultural concept that risks becoming indistinct and often loses its expressiveness as a consequence. And yet, in the context of an economic liberal understanding in which progress and innovation are supposed to be the engines, a culture is born and reflected in the modern entrepreneurial scene. Thus, it is necessary to grasp this culture in its form. Its emotional forces to get to know its moral economy (Daston, 1995: 24) because '[...] moral economies are historically created, modified, and destroyed; enforced by culture rather than nature and therefore both mutable and violable; and integral to scientific ways of knowing (Daston, 1995: 7).' This is why I will examine the cultures around the concept of innovation and their characteristics to approach a culture that conveys something about such ways of knowing.

By way of example, I picked out two descriptions of a 'successful corporate culture' to examine them more closely. Suppose one asks Brian Chesky, CEO of the company *Airbnb*, rather than a cultural scientist, what culture is. The answer sounds quite simple: 'Culture is simply a shared way of doing something with *passion*' (Chesky, 2018: 76). This description becomes even more passionate when reading the brief contribution of the company's head in *The Guru Book*, a guidebook and a collection of experiences of various CEOs of Western countries:

> The thing that will endure for 100 years, the way it has for most 100-year companies, is the culture. The culture is what creates the foundation for all future innovations. If you break the culture, you break the machine that creates your products (Chesky, 2018: 76).

These 100 years do not refer to a biblical revelation, nor do they claim to change, which most cultures entail alongside continuities. Yet, it becomes clear how highly the concept of culture is valued, and its context is associated with innovation and values. More generally, by culture, Chesky refers to what social scientists understand by the term 'habitus' and what Durkheim would attribute to a 'conscious collective'. Through the habitus, socialisation and the understanding of norms become evident. Following the CEO, this becomes apparent in hiring people, writing an e-mail, and walking along the corridor (Chesky, 2018: 76). What exactly he means by this remains vague, and yet there is a hint of some fluid knowledge (Star, 1992) that is supposed to refer to a culture: small process structures and micromanagement reduce the potential for autonomy and, thus, *trust*, he writes. In addition to values that are to be shared, trust is to be created and maintained.

Trust, it is noticeable, is a frequently used term in this context, especially when CEOs describe their company's atmosphere, which is why it particularly attracts my attention. Studies in anthropology suggest that trust is the balancing factor for uncertainty where solid knowledge is lacking (e.g. Strathern, 2005). Tim Ingold, for example, describes its essence as '[...] a peculiar combination of autonomy and dependency [...]. Trust [...] always involves an element of risk – the risk that the other on whose actions I depend, but which I cannot in any way control, may act contrary to my expectations' (Ingold, 2000: 69–70). And yet, at the same time, trust here seems to be one of the best arguments to respond to secrecy, uncertainty, and risk (Corsín Jiménez, 2011: 192). Chesky's text says that only where there is little trust would many rules be needed to compensate for the same (Chesky, 2018: 77). And apparently, this argument promotes the use of trust as a (new) organisational category (Corsín Jiménez, 2011: 178). However, as we encounter it in Chesky's text, the concept of trust does not mean an emotive category but a cognitive one. 'We have accounts of trust as […] a dynamic of "encapsulated interests", where trust emerges as a mutual co-implication of interests on all transacting parties (Corsín Jiménez, 2011: 178).' The alleged trust is supposed to reveal a relationship, reflecting transparency. Everything around the visible thus creates a counterbalance to obligations of secrecy, discretion, and risk. Although accountability is present, the suggested radical visibility blurs it. Therefore, trust only exists in a system that has demonstrated trustworthiness, founded on the flow of information and a solid understanding of the system. This understanding, however, completely contradicts trust as an emotive and creates misunderstandings. Because '[any] attempt to impose a response, to lay down conditions or obligations that the other is bound to follow, would represent a betrayal of trust and a negation of the relationship' (Ingold, 2000: 70). In addition, other emotive illustrations aim equally to bolster the image of trust.

Another CEO, Tine Thygesen of *Mesh*, a start-up network, refers to *loyalty* associated with trust that builds a company's culture (Thygesen, 2018). She also emphasises empowering and challenging employees to get the most out of them. Thygesen repeatedly refers to *humility* and *humanity*, which are necessary to run a company. *Drive* and *passion*, she also notes, are indispensable for a culture that everyone supports. In this respect, it is also important that employees feel this drive and passion for changing things. She writes: 'The start-ups that manage to articulate this clearly can create an almost cult-like atmosphere where the company becomes a major part of the employees' and founders' self-image' (Thygesen, 2018: 79). Quite apart from the fact that here again, the fetishisation of labour, of the product and the self, come into play, the company, in this instance, clearly deals with emotive nouns although it might imply otherwise. According to the statements, culture – here alone, emotional-individual characteristics are included – based on trust, loyalty, humility, humanity, drive, and passion. Depending on which CEO one would ask, there might be one or two more descriptions, albeit equally charged.

The definitions of culture refer to an expanded concept of culture, which combines an open and closed understanding of the concept. It is open due to its flexibility, dynamics, and cohesion (e.g. Bolten, 2007; Hofstede & Hofstede, 2001). At the same time, these cultures try to distinguish themselves from other companies, develop an identity, and 'have' a culture, which in turn points to a closed understanding.

Figure 3: The Emotive Corporate Culture

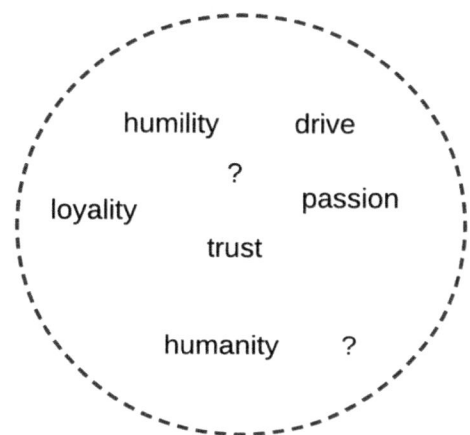

This general understanding of culture was transferred from anthropology to organisational theory (Smircich, 1983) and developed in the early 1980s. Following the critique of the concept of culture, from then on, a new cultural concept of corporate culture is also subject to a vague, broad understanding that is interpreted and treated differently by its users (Smircich, 1983). However, one can determine the term's emotional connotation, interpretation, and use. Even more, it can be said that the concept of corporate culture is quite deliberately emotionalised. The nouns listed are emotive resonators that users can interpret differently. Regardless of the interpretation, they convey a feeling of collective belonging and family, which can be enhanced even further when this (corporate) family becomes a private sphere for the individual and is thus interchangeable with what could actually stand in its place. However, this happens for corporate purposes (e.g. Corsín Jiménez, 2011; Illouz, 2007). After all, what happens when my trust, humility, or humanity comes to nothing? What if trust becomes entirely elusive because a company inherently operates on its own logic, prioritising market interests over the collective's need to interpret and establish a social reference independently? The emphasised terms gain a shared meaning through collective interpretation within a group and are not subject to dictation by a higher authority. It is something learned and handed down and only persists through common culturalisation based on general acceptance and not on being dictated to or postulated by the CEO—terms imposed from the outside degenerate into empty phrases.

Still, the concept of culture is applied because a company, an organisation without rules, does not exist in practice. Although not openly communicated, the concept of culture substitutes what others term rules and a group possessing an evolved culture exhibit typical interaction forms and cultivates diverse practices and rituals.

In this respect, the CEOs' previous statements appear arbitrary, thus promoting certain emotions while tempering others. A corporate culture communicates what is considered adequate and inadequate. In general, culture always refers to the 'how' (Bright & Parkin, 1997: 13). However, how does it occur that—whatever understanding we come across—the legitimisation of the content of culture emerges?

At this point, I refer to the empirical part concerning creativity dispositifs and calls for more innovation, which also results in *a* 'culture' in terms of the ways of dealing with the phenomenon of innovation. Understanding the values and practices a lab or a start-up represents requires knowledge of how it is embedded and financed and the networks it utilises. Equally, it concerns the following questions: How does a company justify its work and product? Under what circumstances was it developed? Who evaluates the product, and who is responsible for it? Chapters V, VI and VII will address these questions, which will undoubtedly provide further research scope.

Observing these aspects, such as (1) the group or actors that are involved, (2) their policies and politics, and finally, (3) their values, tells us a lot about the innovation culture. Therefore, we learn about a group's or society's emotional position and understanding of values. This classification helps determine trends and qualitative indicators of success or failure in innovation-making. These relations find their place on a micro-level (culture). However, they intertwine with national structures and their intermediaries, thus connecting the meso-level and the macro level, such as the European Commission's call for a Union for Innovation. This given structure means that it is observable how nations deal with a global postulate of progress imperatives and if and how this finds expression in economic efforts.

Being primarily an open-ended process, valuing the qualitative exploration of these developments proves significant because, unlike in economics papers, the qualitative inquiry does not presuppose quantifiability in advance. Instead, as in this case, it examines the process's openness, which cannot be planned (Briken, 2006).[1]

---

[1] In this context, it has become clear that empirical attempts to predict success based on quantitative models, such as 'Linear Structural Relations' (LISREL) or similar models, are not particularly meaningful. In this respect, moving away from an R^2 factor is advisable (e.g. Curnow & Moring, 1968; Panne et al., 2003; Roure & Keeley, 1990).

It needs legitimacy for a concept like innovation to become a dispositif and universal recipe for success[2]. Not only does the notion need broad recognition, but equally, what emerges from it, i.e. its ideas, inventions and, ultimately, products. Without its emerging profitable products, the concept would remain empty and fail. Consequently, the concept – especially as its success is not always immediately apparent – needs strong belief (Deutschmann, 2020; see also James 1909; Latour, 1996; Latour, 2010). Therefore, the label and concept offer a frame and support a secure environment for testbeds in the form of 'labs' (see subchapter 4.1.1 and later 5.1). These are quasi-sacred spaces that, similar to religious contexts, offer separate protection for the practice of doing, in this case, innovating. Other texts on innovation spaces also reach for similar outcomes. In the text 'Innovation Spaces' by Moultrie et al. (2007), the authors conclude that more creativity results from the space created and identify various factors that are (supposed to be) promoted by the space.

On the one hand, Moultrie and his colleagues define *competitiveness* as a strategic goal for companies, aiming to reduce costs and enhance employee productivity. Moreover, they seek to enhance the quality and quantity of ideas while promoting teamwork through improved communication structures and closer collaboration within the lab. Furthermore, the quality and quantity of ideas should be improved. Promoting the ability to work in a team should involve improving communication structures and enabling closer collaboration within the lab. In addition, the option of 'customer input' plays a role, i.e. the opportunity to receive ideas from outside and, in general, implement specific skills should be provided by installing the creative lab.

On the other hand, the authors also acknowledge the *symbolic power* of the lab, which I believe is manifested in all the mentioned factors. It is not only about the strategy and cultural incorporation of the company but also about the corporate values conveyed by the facility (see: Moultrie et al., 2007: 57). Therefore, while new structures emerge as described above through the creative space, a new working culture simultaneously arises, which is inscribed and communicated. Ultimately, legitimacy stems from establishing these labs' from above,' thereby ensuring their favourable reception under the mediation of modern and open structures. Different forms of dealing with innovation, specifically innovation cultures – whether nationally generated or institutionally cultivated, as in companies – can be identified. A description may read as follows:

---

2   From Latin 'lēgitimus'/ 'lēx' meaning law: in this context, it means a set-up through a 'law' or rule, rather than a dispositif; a dispositif that describes what is expected or right to do and act in society.

Reaching innovation is a key challenge for any business in a competitive market. However, often the best source of innovation is actually within the company itself – the employees. The most successful companies are the ones who [sic] capitalize on this asset and create a culture of innovation, using employee suggestion software to transform ideas into results (Qmarkets, 2021).

Furthermore, the ability to innovate is becoming a hallmark of these very societies that are beginning to legitimise their work through the postulates, demands, and credos just mentioned, which, in part, remain unquestioned (e.g. John, 2012; Latour, 1996: 287). It is neo-liberal and forthcoming structures, as in organised innovation-making such as 'labs' that lead to an understanding of 'a generalised perception or assumption that the actions of an entity are desirable, proper, or appropriate within some socially constructed system of norms, values, beliefs, and definitions' (Aghamanoukjan, 2012; Suchman, 1995: 574). On the one hand, the inflationary use of the word may create an inherent logic of legitimacy, but on the other hand, on closer inspection, it behaves just as insubstantially. The fact that a term now replaces many others, such as 'novelty', 'discovery', or 'improvement', does not generally imply a better understanding or method for how this content comes about. The arbitrariness that pervades this triumph ultimately points to the non-verifiability of the concept since it does not have any standard criteria of legitimacy and quality.

4.1.3 Expectations as Iterations in a Black-Boxed System

Theoretical solutions to problems are, by their nature, promising. Until implementation, they do not need to prove their functionality, but in the process, they serve as a canvas for all hopes and unmet expectations. In this respect, in theory, they provide a space for all actors and their expectations. As mentioned in the previous chapter, the future is only the servant of a failed past in the present. The promises represent solutions in the future to problems from the past that emerge in the present. Therefore, the future as a period is not independent, but through its temporality, it is always dependent on what has gone before. Thus, the expectations, whose origins are previously identified, also rely on this temporal sequence. In the process, a remarkable degree of promise arises precisely through creativity. The diversity of ideas leads to the will to experience the *eureka* moment. However, the nodes of the iteration loops are significant, i.e. those points where a repetition loop, a renewal of the idea, a 'new start', and suchlike commences. They are where either new actors join, new proposals are made, and/or previous ideas are rejected and represent the 'grinding points' of an idea toward the final product.

Figure 4: Iteration Loops of Expectations During the Development Process

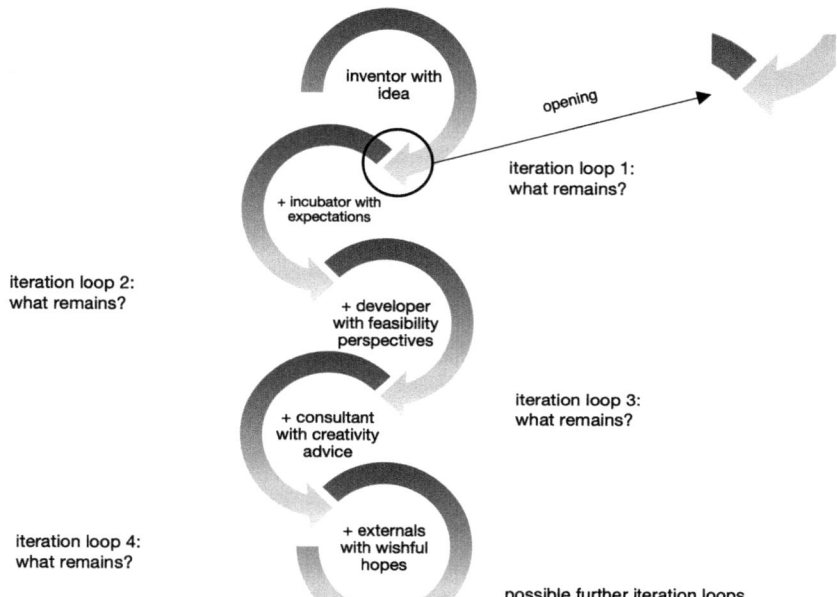

Figure 4 illustrates the iterative nature of the development process of a prototype, often referred to as iteration loops. Moments of *opening* and *closing* of the prototype characterise these loops (Corsín Jiménez, 2014; Dickel, 2019; Guggenheim, 2010, 2014). Opening signifies the opportunity to incorporate new ideas and make changes, followed by a moment of closure. These phases alternate, highlighting the continuous adaption aspect of the prototyping process. Moments of opening appear, as indicated here, by opening a new circle (see enlargement in Figure 4), for example, when new actors join a team, after team meetings and consultations, or because the milestone plan requires it. This open space indicates utopia in its original sense – a 'non-place' (from ancient Greek οὐ ou 'not' and τόπος tópos 'place'). The idea space 'topos uranios' is the heavenly place where the idea dwells (Bloch, 1980: 43). It is not a geographical 'place' but a space – perhaps even just an 'aching gap' (subchapter 3.4.1) as we see here – that represents the open portal to new creations: *innovation*.

As the term implies, moments of closure entail the conclusion of opening processes, wherein participating actors decide following the presentation of proposals or improvements. A decision (temporarily) closes the respective opening process, and the two processes alternate until completing the final product. These moments of openings and closings thus correspond with changing and defining the design of a prototype. The prototype, due to its experimental character, indeed allows for failures. However, it also enables a democratic and open form of technology development, at least in theory or with concepts such as CTA (Bowman et al., 2017; Rip et al., 1995). This approach not only involves the participating actors, as in the case of Susan Leigh Star's and James Griesemer's 'Boundary Object', which is fed or interpreted by information from the outside but retains the identity-forming part that becomes the 'core of the thing' (Star & Griesemer, 1989). It also creates a user-oriented design, which is often called 'open innovation' (Corsín Jiménez, 2014: 382). Therefore, the prototype is the materialisation of many expectations and the result of an experimental process in which decisions or consensus are reached in a communicative-collaborative – at best democratic – manner. Inscribed in them, we find with each narrative an expectation, a hope, an attitude and stance, an opinion, and ultimately emotions. As long as the technological artefact is a prototype, this is inevitable; it is a product of cultural discourses, an object of socialities and relations (Law & Mol, 1995).

However, even if the appearance of prototyping seems to be a democratic process, the overall opportunity for technology development remains a black box (Collins & Pinch, 2014; Latour, 1987). In moments of technology and innovation development, the information does not circulate unhindered as it does not recur to materiality and expected effectiveness. However, it is much more socially or communicatively conceived (Reinhart, 2016: 166). Adding to the complication, technical developments, although portrayed as such, cannot be convincingly interpreted as technical innovations developed in response to a problem (Reinhart, 2016: 166). Thus, the development of a technological artefact intended as an innovation presents a paradox. Although different actors with different backgrounds come together in so-called incubators to work on a project and try to reach an understanding and agreement in the process of working together, they are obliged by the fragility and uncertainty of the development process to maintain silence. This way of collaboration seems typical among incubators and makerspaces, ensuring that there is neither idea theft nor too much (or unintended) leaked information. The incubator must serve as a sheltered space, meticulously crafting an optimal environment for teams to innovate (albeit artificial and constructed). Within this context, stakeholders must ensure that knowledge dissemination is prevented until it is deemed reliable and secure, as demonstrated in the empirical findings (see Chapter VII).

These incubators bear a resemblance to laboratories, which '[are] the result of a procedure that separates between an outside, an environment that is considered negligible for some epistemic claim or technological invention, and an inside, a (partly) controlled environment that is considered relevant for this claim or invention' (Guggenheim, 2012: 101). They exist in a state of partial seclusion and isolation, endeavouring to address a commonplace issue that is incongruous with their daily existence within the incubator. The inventors transpose a problem from a 'real world' context – potentially their own – to be examined under microcosmic conditions. Their sociality with each other within the emerging team and with the object (Knorr-Cetina, 1997), as well as the settings and prerequisites, are constructed. It is imperative to extricate the problem from its initial context to facilitate a thorough examination and derive a viable resolution. The problem at hand, and ideally the resolution as well, necessitates the attainment of control.

Moreover, one can introduce alterations, encompassing potential future scenarios that influence the utilisation or conduct of the subject matter. Ultimately, one can modify the laboratory environment to restructure it distinctively and conduct experiments. The capacity to isolate, regulate, and manipulate epitomises the essence of the laboratory condition. Nonetheless, while isolation and controllability are indispensable for innovation studies, they also possess adverse facets.

While these makerspaces, incubators, and living labs rely on their confidentiality clauses and their keyword *innovation*, which guarantees a cloak of silence, the problem arises that knowledge generation generally remains in a black box. Developers and incubators do not disseminate this knowledge; they do not practice openness, and, ultimately, they seem to deliberately delay the publication of findings on innovation research and development (R&D) despite technological advancements in the field (Cristea et al., 2019; Ioannidis, 2015). In this way, innovation spaces operate secretly and create a space that offers the necessary isolation to innovate competitively. However, due to the constant unity, there is the danger of verifiably being contrary to constructive technological development. Eventually, verifiability does not only mean an alternative form of technology construction that develops in the interest of society, as the CTA proposes. In addition, verifiability refers to the legitimate interest in innovators' capabilities and whether they can ultimately keep their promises. If the sense of responsibility for technology development and the resulting ex-ante promises were to be lost, this would not only be questionable for moral reasons, but innovation development would consequently become obsolete and abolish itself.

4.2 Structures of Innovation II: Narratives, Myths, and Beliefs

Telling stories within groups and societies is considered original and natural. It is an informal and necessary function of being human. Storytelling is the verbal expression of the imagination of images, our consciousness, and what we actively perceive (see: Comer & Taggart, 2020: 25). Stories stir something in us; they speak to and touch us emotionally. We convey and control what we want to say (and how) with words and use stories as an instrument to reveal what is on the inside and persuade the other party. Therefore, for entrepreneurs and innovation teams, much depends on the narrative surrounding a prototype, its company, or the team members – for example, their legitimacy, financing, or the team's productivity.

Although the field of storytelling, narratives, and communication systems is well-researched in the humanities and social sciences (e.g. Bausinger, 1958, 2016; Comanducci & Wilkinson, 2019; Friedl, 2013; Ricœur, 1988, 1995), as well as psychology (e.g. Comer & Taggart, 2020; King, 2000; Smorti, 2020), there is a relatively large research gap for the field of entrepreneurs and the start-up scene in general (Borghoff, 2018). However, this field is particularly revealing when investigating motives, strategies, and the so-called 'gut feeling' for a 'good' investment (Villanueva, 2012: i.a. 38).

As in the example of *Theranos* summarised at the beginning of this chapter, the narratives of innovation are the easiest to analyse over time and from their end as they expose their adaptations. The narrative or founding story has many functions and is usually highly emotionally charged. As described under the aspect of legitimacy to innovation, too much depends on imagination, belief, and interpretation. Since these are communication mechanisms in the constructed system of innovation, Luhmann's concept of a social system is correct at this point (Luhmann et al., 2013; Müller, 2013). '[…] Stories that are told about the system outside of it, in its environment, and that is, so to speak, processed, modified, adapted or rejected in the system as an intervention, are meaningful for the identity of a social system' (Müller, 2013: 139). This chapter, therefore, analyses how narratives evolve, how they are emotionally constructed, and why they may change over time.

The narrative functions represent identity-forming elements for a group or, generally, for the people outside of it. Further, the 'good story' related to innovation, together with imaginatively linked images, creates a general acceptance in society and gains access to financial resources and networks. Finally, it is a roadmap for the developed prototype, and through what it conveys, it connects the past with the present and future. Furthermore, this story is also reflected in the materialisation of the prototype and, later, the final product.

4.2.1 The Evolving Narrative on Innovation

For innovations, or what is called innovations, it is often factual that a problem precedes the original idea (see Chapter III). As Chapter II described, a conscious experience is often a clue to what confounds, challenges, disturbs, and displeases. This problematic starting point ultimately enables one to imagine an improved state. The potentially improved but still imagined state is thus the starting point for innovation and is frequently described in a revolutionary and emotional fashion as outlined below:

I grew up in a family that was very focused on the belief that we are all here for a reason and try to make this world a better place. And that we have a purpose. And I thought I was going to do what my dad did, which was work in disaster relief. Because I grew up in a house where I was surrounded by pictures of him helping people when really bad things happened. And over time, I started to see business as [a] vehicle for making a change in the world because you have total control over what you decide to do and how you decide to do it. [...] When I spent time thinking about what was the most valuable thing that I could do with my life, to me, there is nothing more valuable than being able to change the reality in our world, which is that all too often, people we love are lost because you find out too late in the disease progression process to be able to do anything about it. And the fact that making laboratory testing more accessible is a way to help change that. And that is part of getting rid of that big bad needle (Computer History Museum, 2014: Interview with Elizabeth Holmes, CEO of Theranos, at December 9th 2014 led by Michael Krasny).

Often, a detected *problem* or a *sense of injustice* precedes the revolution narrative. In the example here, it is not just the 'big bad needle' that was, and as Elisabeth Holmes mentions in the interview, still is, terrifying for her. She also feels the urge to make a difference or bring about change because she grew up surrounded by people who exemplified the importance of making a difference in the world and for one's life to be meaningful. Often, individuals mention that the initial desire to effect a change has either been present for a considerable duration or has suddenly emerged due to an enlightening everyday situation. Therefore, narratives that change the world require a foundation that sets the stage for their actions and motives. Society often highly values and thus utilises meaningfulness and significance (Graeber, 2018). Hence, employing these terms justifies one's actions or the desire for societal and personal change. One often equates meaningfulness with the value of one's work and self. In addition, the idea and the company are valued extraordinarily highly through social recognition and what is considered morally acceptable actions. The (linear) narratives often follow the pattern:

Figure 5: The (Linear) Narration Pattern

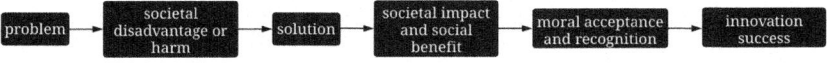

The problem refers to a social disadvantage or even harm; the latter mainly counts for health innovations. The solution has a societal impact in the sense that it benefits everyone. If this is the case, the innovation is morally accepted and recognised. In sum, innovation is likely to be a success. According to this, problem-based creativity applies in most innovation narratives, as shown later again (see Chapters V and VI). Thus, a *creativity dispositif* (Reckwitz, 2017) based on discovering the problem returns to the way of innovating. In this respect, it becomes a race among those who want to be creative to discover a problem that a solution can follow.

Hence, Reckwitz's thesis remains partially valid:

> The regime of novelty produced by the creativity dispositif in all its parts is founded on novelty as a stimulus. What counts for it is the production and reception of constantly new stimulus events, which should be as intense as possible, and the interest of which lies in their immediate presence. The aim is not to be better but to be different (Reckwitz, 2017: 666).

Importantly, in this context, novelty does not necessarily provide stimulation. Instead, *the discovery of the problem* acts as the catalyst for the emergence of new things. While creativity certainly maintains its *dispositif*, its origin shifts. It is not about creating novelty at all costs but about the crucial process of problem discovery. This process, where modern humans stay true to two of their abilities: the conscious act of creativity and the awareness of the problem that makes them discoverers, is what truly engages us in the discussion of creativity.

4.2.2 Mythmaking, Belief, and Performance

Myths are part of every society. As narratives, they are political instruments of identity formation that can promote a sense of belonging. They structure the past (Münkler, 2009) and serve as moral guidance in the present and future. This subchapter shows how mythmaking becomes relevant to innovation as it is more than mere founding motives shared at congresses and funding platforms since myths equally convey the team's emotional values that are (supposed to be) shared. They are the glue of the confessional community and serve to convince the outside world. 'Myth is depoliticized speech' (Barthes & Lavers, 1972: 142) as an achieved effect through language and rules that resembles a message. The myth has a system in its form of communication, and because it is so easy to integrate into everyday life, everything can be (and become) a myth. The condition is suggestive, states Barthes, and he is correct in this view (Barthes & Lavers, 1972: 142), for indeed, the mere description of an idea's form, of what it is supposed to do, does not apply to its publicity. It is not the descriptions that adorn it. An idea, a potential innovation, lives from itself, its projection, and the stories about it. It is not the mere words that give it a character, but what an individual associates with it that is emotionally charged. It lives from society's consumption, which needs to be developed and satisfied, and from images filled with meaning; it lives from semiotics. Nonetheless, it can also perish when society exposes the images as deceptive, when they no longer meet societal needs, and when they fail to evoke any emotion whatsoever.

In this context, people often perceive the creation and transmission of myths as something prehistoric or pre-modern. They associate it with a time before the Enlightenment; thus, it no longer aligns with the logic of modern rationalities. Yet, these myths are inherent elements of society that we cannot eliminate. They bear witness to our emotional interpretation and assimilation of our experiences and socialisation. They represent the very essence of our culture. We constantly interact with the myth; we apply it to our face when we look in the mirror, we drive it, and we type our messages on it. Furthermore, just as with products from advertising, every innovation comes equipped with it. Therefore, every innovation that presents itself as a 'superlative object' has, in some way, 'fallen from the sky'(Barthes & Lavers, 1972: 88).

> We must not forget that an object is the best messenger of a world above that of nature: one can easily see in an object at once a perfection and an absence of origin, a closure and a brilliance, a transformation of life into matter (matter is much more magical than life), and in a word a silence which belongs to the realm of fairy-tales. [...] [Those are] objects from another universe which have supplied fuel for the neomania of the eighteenth century and that of our own science-fiction (Barthes & Lavers, 1972: 88).

Whether it is the new Citroen, as Roland Barthes suggests, or any other novel phenomenon that claims indispensability, the narratives at play are interchangeable, flexible, and potent. They do not just refer to our desires but can create them, setting the stage for their own satisfaction. Consider a random example from the *Kick-Starter* platform to understand the role of narratives and myths in crowdfunding. Here, *Indiegogo* or *Kick-Starter* Projects present their ideas and prototypes, aiming to collect funds from potentially interested individuals. The founders then invest the raised money to transition the prototypes into production. They offer early customers various donation variants, each with its own set of perks. A small donation is equivalent to a 'handshake' (sic!), but a larger investment could secure a first version of the series product upon its earliest release. In this way, the 'packages' available for purchase expand in value with the donation size.

The following example is a variant of a toothbrush – a start-up idea from Austria dating back to 2016. The company started their funding period on Indiegogo and Kick-Starter in the summer of 2017. The idea was a fully automated toothbrush that cleaned your teeth within a few seconds. It looks like a pacifier from the outside with a protrusion for the teeth. This new type of toothbrush does not require a hand to hold it while it scrubs; it is a device placed on the teeth, which then cleans them.[3]

> Amabrush – The 10-Second Toothbrush (Kick-Starter-Project)
> Do you like brushing your teeth? Especially at night when you get ready for bed? Amabrush is the first toothbrush, that cleans all your teeth at once in just 10 seconds! You never have to brush your teeth again!
> Amabrush is the world's first, fully automatic toothbrush. This patent-pending device brushes all your teeth at once, [is] fully automatic, and finishes in just ten seconds. All you have to do is press a single button, wait ten seconds, and you'll have perfectly clean teeth!
> And further: Let's face it: brushing your teeth is not exactly the sexiest thing on Earth. You have to squeeze, scrub, gargle, spit, rinse and floss every morning and evening, every day of your life. Many of us hate brushing our teeth so much that we avoid doing so whenever possible—even though we know we shouldn't... Brushing our teeth at least twice a day maintains good dental health. This is why we invented Amabrush—a device three years in the making with a single goal: to make toothbrushing quicker, automatic, and more efficient so you have more time for the relevant things in your life (Amabrush, 2019).

The product represents the facilitation of everyday life, a healthier version of what we know, but in a more convenient and user-friendly format. The advertising texts often agree on this, and they, too, become interchangeable.

We find words such as *just, at once, never, fully automatic, all you have to do, a single button, and perfectly clean*. The second part continues with the negative aspects: *not the sexiest, every day of your life, hate, avoid*. And again, the single goal: *quicker, automatic, efficient – to have time for the relevant things in your life*.

At this juncture, the intended achievement of language becomes clear. Initially, certain words aim to communicate a specific ease. A unique excitement seeks to present a modern approach to one of the most routine daily activities: brushing one's teeth. The revolution of even the simplest things carries a sense of 'extraordinary simplicity'. The suggestion projects an image of a simpler, better life devoid of effort and expense, even though it pertains to a cause of such importance as the company declares. Despite the apparent neglect of the practice of brushing your teeth, it remains essential for health. This ambiguity often arises; it is the interplay between

---

3   The start-up 'Amabrush' met its demise in 2019. Despite a successful prototype, the Austrian company failed to deliver its promise of a toothbrush that could produce clean teeth. This failure led to several fraud allegations and left several thousand customers fuming. In a disappointing turn of events, the fraud case against the manufacturer was dropped in 2020, leaving the deceived customers without compensation. A staggering seven million euros had been collected via crowdfunding platforms, making the failure all the more significant.

relief and necessity, whether for one's own body while driving or in the household. In all innovation sites, one can find pairs of opposites and antitheses, which become compatible and harmonised through the expanded possibilities of innovation.

These myths, prevalent in 'hot' or 'heated' societies or cultures (Assmann, 2011; Levi-Strauss, 2021), embody a unique complexity. They are the products of flexible societies that, while desiring change and evolutionary progress, remain rooted in the narrative form of mythmaking. Suppose we start from Levi-Strauss's and Jan Assmann's thesis of hot cultures, societies and cultures driven by progress and expand on Ulrich Beck's thesis that this very society is transitioning from the contours of industrial society to the 'risk society' (Beck, 1986) where a sense of threat is pervasive, albeit sometimes subtle (Beck, 1986: e.g. 59) or, at times, unconscious.

At this point, we must question the extent to which the drive for renewal, improvement, and innovation stems from the uncertainty inherent in these hot risk societies. Could it be that these *societies*, feeling insecure, seek to compensate through adaptation and change, namely technological progress? Following this line of thought, a state of uncertainty likely always precedes the status quo within a society and the status of its technical development. This state is perpetual, as the present, being uncertain, finds its resolution in the future. However, the future is merely a reflection of the present and is, therefore, also uncertain.

And yet, the same applies to the innovators themselves, who want to enter the incubator and grow in it. Entre- or intrapreneur (Parker, 2011), inventor, and innovator are their names, and they establish their own creeds. They exude a profound self-belief in themselves, their idea, the product, and the team; they believe in the consumers who will (should) discover the product's intrinsic value. This unwavering belief is the foundation of their trust-building process. And vice versa: the others, e.g. the investors and customers, are inspired to believe in them, the idea, the product, and the team (Villanueva, 2012: 136 f.). The investors and customers need to think that the money is well spent, that the idea can succeed, that the right people are working on it and that suitable suppliers have been chosen.

Religious analogies are not unusual, particularly when discussing meaningfulness, faith, and impact. A 'cult-like' atmosphere, as alluded to in subchapter 4.1.2, suits the insecurity and exposes even more in this regard, as we can read in Georg Simmel's 'A Contribution to the Sociology of Religion':

> All religion contains a peculiar admixture of unselfish surrender and fervent desire, of humility and exaltation, of sensual concreteness and spiritual abstraction, which occasion a certain degree of emotional tension, a specific ardor and certainty of the subjective conditions, an inclusion of the subject experiencing them in a higher order – an order which is at the same time felt to be something subjective and personal (Simmel, 1905: 362).

In the preceding descriptions, religious acts are paralleled to compensate for the uncertainty and lack of order and knowledge, and this order is subsequently found in teams, the incubator, and in this 'trust from above', which is supposed to provide support. In this regard, the existence of certainty underscores the crucial role of trust in forming knowledge. This trust enables inventors to perform with belief and conviction (Seidenschnur, 2019). They perform on demo days; they shine, and they are convinced. They perform at TED (short for: Technology, Entertainment, Design) Talks and earn applause. They are performers because they must convince others.

With failed companies, such as *Theranos* or *Amabrush*, the investors' shock suddenly becomes the unmasked naivety that they 'want[ed] to believe' (Yahoo Finance, 2021 Documentary: 57'). The interviewees talk about 'how much they wanted it' – how much they wanted to see the idea succeed, upon which the experts join in and recite the motto: 'Fake it, 'til you make it'. The myth is an aid to one's own faith relationship with oneself (Latour, 2010). It drives the performance of the self-confident founder who knows how to convince the people around him.

4.2.3 How Narratives Adapt

Narratives may also adapt during prototype development. First of all, there is the discrepancy whereby a single person, the creator, initially has an entirely different idea of the possible development of a product in mind. Hence, as already discussed under the aspect of experience and the moral economy (subchapter 3.4), an initial concrete idea is associated and conveyed with an experience and a problem, both of which result in a narrative. As more team members join the initiative, these ideas and associated images add up or are reduced, similar to the iteration loops of a prototype: they overlap, reinforce, or partially exclude each other, weaken, and renew. Like the prototype, its founding narrative is also subject to a grinding process, i.e. various adjustment processes that are decided on at different moments alone or together in the team. However, it remains an ambivalent process that can be connected to several insecurities, as Caroline Bartel and Raghu Garud pointed out:

> [B]ringing people with disparate perspectives and capabilities together during the innovation process can, in turn, create other difficulties. For example, ideas that come from different parts of the organization may remain underused to the extent that people are unable to see their relevance to their own work. Also, dysfunctional confrontation can arise as people with diverse backgrounds and expertise interact, thereby undermining innovation. Such unproductive social interactions can exacerbate the uncertainties inherent in innovation processes and increase the chances of generating suboptimal outcomes (Bartel & Garud, 2009: 107).

In this respect, a shared corporate culture can help develop shared values, norms, and beliefs that allow a common ground for social action during the innovation process (Bartel & Garud, 2009; Jelinek & Schoonhoven, 1990). On the other hand, these cultural structures can also severely affect employees, leading to contrary developments. According to Bartel and Garud, negative stress can arise due to different working methods, leading to tension (Bartel & Garud, 2009: 108). At this point, the innovation narrative can assist and become the mechanism that enables both coherence within the team and flexibility for the people involved. Hence, the interconnections between narrative and corporate culture become evident.

Consequently, for tactical reasons, a narrative is often reduced in some respects and enriched later, for example, by a plot. This strategy typically involves a conscious process where a decision is made to disclose (or withhold) information deliberately. This decision hinges on whether sharing some information might be beneficial or if the situation is too delicate to divulge. This decision always relies on the estimation of potential future success.

Equally, unconscious moments occur; at times, one might inadvertently omit something, or a certain aspect might take precedence over another due to its current relevance. Nonetheless, it is essential to note that the narrative continues to exert control. Further, one can find guidance websites for entrepreneurs and teams on the importance of storytelling and how it can create a certain legitimacy (e.g. Day & Shea, 2018: "Grow Faster by Changing your Innovation Narrative"). Among other things, there are references to economic aspects that promise to grow faster through a better narrative. The MIT-Sloan Website states that:

> An innovation narrative is an oft-overlooked facet of organizational culture that encapsulates employees' beliefs about a company's ability to innovate. It serves as a powerful motivator of action or inaction. We find innovation narratives in two basic flavors: growth-affirming and growth-denying, or some combination thereof (Day & Shea, 2018).

In addition to widely touted success concepts such as Innovation Boot Camps or Design Thinking, the magazine article concludes that the narrative is an often-underestimated factor. As previously mentioned, the narrative can serve as the cornerstone of the development and success strategy, acting as a mechanism in challenging situations where translation is necessary to establish legitimacy, even within an organisation (Bartel & Garud, 2009). Translation involves tailoring the narrative to the audience and adapting it to the various actors involved in the process of innovation communication (Latour, 1994). At this stage, the narrative must change solely within a team or company to generate and maintain its internal legitimacy and motivation to avoid tension or stress. In addition, this circumstance is equally valid for external parties. Thus, this is not a mechanism to ensure coherence and flexibility but to maintain external conviction, legitimacy, and acceptance. Once the team stabilises its identity, it can communicate content externally, where the narratives are adapted and translated. In conveying innovation, another identity moment that promotes acceptance and social capability must be constructed. It becomes clear that the emotionalisation of the narrative plays a recurring role for reasons of identity and empathy, both within and outside the team (Villanueva, 2012).

Consequently, there are connections between narrative and growth, emotionalisation and legitimisation, and further ethical aspects are linked. There are plenty of websites providing advice and service to founders to point out that certain target groups are more likely to be addressed by exact 'wording' and that a business should also convey a specific message through the word, whereby, above all, social norms and values should be considered (e.g. N.A., 2022; Williams, 2022). The 'wording' and the set of values vary depending on the product and the target group addressed. One should consider the overall communication method through text design, which equally involves colour, image, and content. This approach creates a resonance space that engages the emotional level.

Interestingly, in this context, one can observe how business narratives are also changing, especially concerning the shifting norms on climate change, sustainability, and social responsibility (Hinkel et al., 2020; Kuenkel, 2018; Mackintosh, 2021). There is a noticeable shift towards new business strategies in the media, literature, and political campaigns. These strategies do not primarily focus on profitable 'how to make money' approaches but rather on sustainable business practices. This shift aligns with the Green Deal initiated by the European Commission in 2019.

Hence, the advice for companies and innovators does not necessarily recommend a rethinking but primarily a retelling. With reference to economics, ethics, monetary incentives, and globalisation, companies are encouraged to revise their approaches as a society or group's value system and emotional attitude continue to steer its economic intentions and, thus, its innovation ventures.

4.3 Values and Evaluation

In the previous two subchapters, 'Structures of Innovation I and II', I discussed the structures of innovation practices, which involved the spatial allocation of practices subsumed under innovation-making, such as makerspaces and innovation labs. Additionally, the emphasis is on culture, specifically, the structures manifested in daily practices and customs and how individuals narrate them. This focus is particularly relevant in terms of the practices and narratives conveyed, that is, the relationship between what people do in their everyday lives and how they discuss it. Concerning how practices form and constitute, our identity determines the value we ascribe to the activity (Krüger & Reinhart, 2017), how we view and value innovation and, ultimately, how we feel about the degree of innovation of an artefact.

The fields of 'valuation' and 'evaluation' have been emerging and developing in sociology for several years by dealing with different phenomena that fall under it (Krüger & Reinhart, 2016; Krüger & Reinhart, 2018). It is either the field of investigation of the attribution of monetary or non-material value to material and immaterial goods in the realm of nature or human life or rankings and ratings as formalised valuation practices (Krüger & Reinhart, 2018: 2). Additionally, the sociology of evaluation contributes to normative value orders as orders of justification (Krüger & Reinhart, 2016: 487).

Generally, visible and invisible evaluation processes can differ in their logic, although they do not necessarily do so. For example, in relation to innovation, there is always the question of whether it is marketable (Reinhart et al., 2019), and hence, market logic comes into play. As shown later in 6.2 or 6.4, for the field of medical technologies, it can be questioned to what extent patient interests differ from these market logics or whether a processual convergence of capitalist market logic and patient interests becomes relevant here.

4.3.1 Constructing Value Consensus

As demonstrated in the preceding chapters, the innovation and prototyping process involves a continuous re-evaluation. The thesis posits that just as expectations and prototypes evolve over time and undergo refining processes, a parallel evaluation of the same occurs. Depending on the observer and the actor, different evaluation logics come into play during the evaluation process, which the prototype must satisfy as a product by the end of its development. The evaluation logic hinges on the individual's position and discipline: their professional background and emotional perspective. Therefore, in the moral economy, not just the expectations, perspectives, and claims converge but also the associated evaluation logic.

Consequently, the prototype depends on these circumstances and all those who negotiate them. Typical questions in this process include: Who gains the upper hand? How do these negotiation processes shape the prototype? Moreover, does the prototype foster a confident expectation through its further potential for possibilities, for instance, with the incubator and the business angel?

Figure 6: Evaluation Interactions Between the Prototype and the Individual

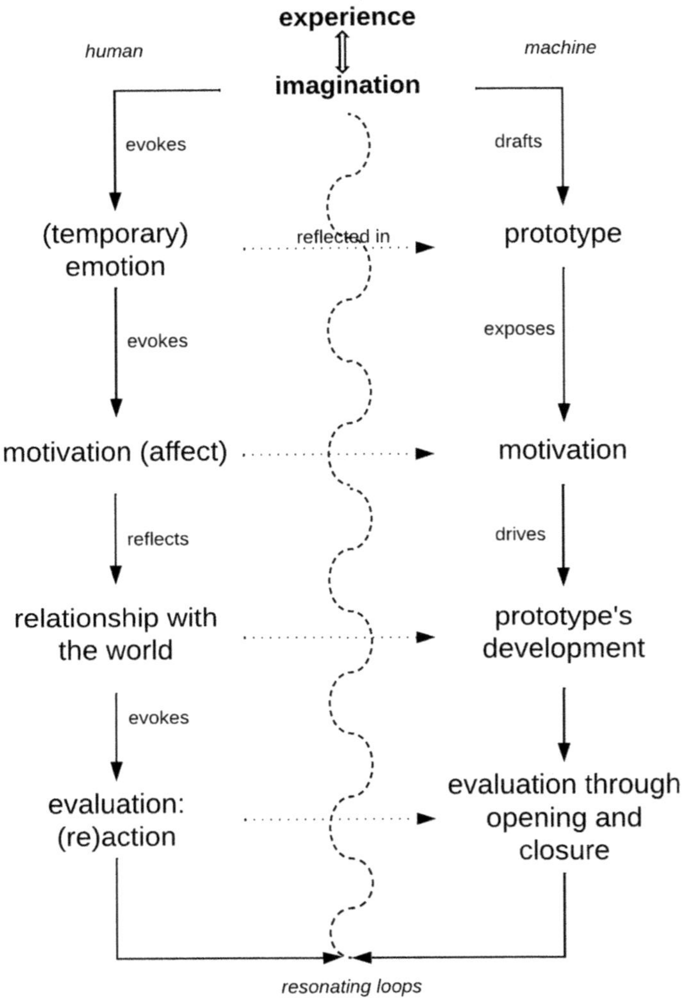

```
                      experience
      human              ⇕              machine
                      imagination
    ┌─────────┐        ╱               ┌─────────┐
    │ evokes  │       ╱                │ drafts  │
    ▼                 ╲                ▼
 (temporary)    reflected in         prototype
   emotion    ·············►
    │                 ╱                   │
    │ evokes          ╲                   │ exposes
    ▼                 ╱                   ▼
 motivation      ·············►        motivation
  (affect)           ╲
    │                 ╱                   │
    │ reflects        ╲                   │ drives
    ▼                 ╱                   ▼
 relationship    ·············►       prototype's
  with the           ╲                 development
   world              ╱                   │
    │                 ╲                   │
    │ evokes          ╱                   ▼
    ▼                 ╲               evaluation through
 evaluation:    ·············►        opening and
  (re)action         ╱                   closure
    │                 ╲                   │
    └─────────────────►◄──────────────────┘
                   resonating loops
```

Figure 6 is a further elaboration of the above sketches, especially that of Figure 2, albeit it means a simplification and does not refer to later ruptures that occur in this process. It shows the parallel course of individual human emotion and evaluation processes and the development of the prototype. Both processes start with the result of the interrelation of experience and imagination, evoking distinct emotions based on previous expectations and accompanying imagination (left). Likewise, one's background, i.e. *a* previous experience, evokes a particular idea of a model that is supposed to address everyday issues (right). Both lead to a specific motivation. The left side illustrates how a specific emotion prompts action, creating a need to alter the problem. On the right side, the associated rationale for developing a specific prototype is displayed. Hence, it is a purposeful motivation justified by a particular problem. The left demonstrates motivation, specifically how emotions guide individuals in their relationship with the world. The behaviour triggered by emotions thus provides insight into how an individual relates to their environment. On the right, this relationship to the world, i.e. the motivation to want to solve a certain problem, provides information on how the prototype develops. This, in turn, reveals the background, motivation, and a particular relationship of trust or acceptance between innovation and society.

Finally, both processes lead to an evaluation process. On the left is the reaction to the prototype, leading to a judgement. On the right is the evaluation process, which affects the prototype's further development. The individual steps resonate with each other and interweave; they behave in an oscillating, and partly circular way, whereby circular in this context means that processes can start from the beginning. The (non-)completion of a product is, in turn, reflected as a new experience in our evaluation patterns in a new development process.

Generally, the processes involved in the constitution of knowledge grapple with the assertion of science's supposed objectivity (see subchapter 3.4). The expectation is that evaluation schemes follow 'objective criteria' and are rationally generated to ensure some level of security (Reinhart et al., 2019). Yet, sociology has demonstrated how these can be deconstructed. Consequently, evaluators apply different forms of evaluation schemes to assess innovation. Quasi-objectified or seemingly objectified evaluations determine whether an idea merits funding, development, marketability, and societal value. These assessments also encompass understanding how we value innovation, i.e. whether society is more likely to accept one idea over another. They also provide insights into the conditions and circumstances under which we develop something and how transparent the development process should be.

Following this, we will explore how society brings specific inventions to life and the crucial elements for an artefact's implementation. We can question whether it is possible to establish criteria to judge the potential success of something. We will examine how we assess the success or failure of an idea and its development and whether we can pinpoint specific reasons for a lack of success. In this context, the question of how we evaluate the innovation processes themselves emerges. We will explore how we confirm an invention's innovative potential. Given the Theranos case and the absence of verifiability, we must critically scrutinise and question the success and transparency of such a process.

In the subsequent discussion, we will consider how we develop specific inventions and the essential factors for an artefact's implementation. We will ponder whether it is possible to establish criteria that we can use to judge whether something will be successful. We will investigate how we evaluate the success or failure of an idea and its development and whether we can explicitly state why something was unsuccessful. Furthermore, the question of how we assess innovation processes themselves arises in this context. We will examine how we verify the innovative potential of an invention. Based on the case of Theranos and the lack of verifiability, we must critically examine and question the success and transparency of such a process.

4.3.2 Serendipity or a Matter of Perspective?

Louis Pasteur once remarked that '[Chance favours the prepared mind]' (Vallery-Radot, 1926: 76). In his inaugural address as newly appointed professor and dean at the new Faculté des Sciences in Lille, France, on 7 December 1854, he questioned the mere coincidence of inventions and discoveries. He thus refers to the urgency of foreknowledge, or an open mind, to come across anything that one might later call chance; a quiet hunch, an open eye for what is happening around one, to see a problem. This assumption is correct insofar as previously described. One cannot comprehend a problem without being alert to one's environment. In the annals of scientific history, the term 'serendipity' frequently appears in these contexts. It denotes accidental discoveries, such as penicillin, X-rays, sticky notes, and the 12 moons of Jupiter discovered in 2018. These discoveries represent findings that ideally would have resulted from deliberate planning or searching. Although the term serendipity first appeared at the end of the 18th century and originally came from a Persian fairy tale, Robert K. Merton introduced it to the social sciences in 1958 with his book 'The Travels and Adventures of Serendipity'. Researching and searching are part of everyday life in science and describe its modern character. Irrespective of whether one considers coincidences, chance discoveries, or luck, one simultaneously walks the fine line between knowing and not knowing one's field or the phenomena taking place in it (Rheinberger, 2014). Hence, by its very nature, research is an activity full of surprises, as knowledge cannot always be located and thus remains unpredictable and, to some extent, constantly an experiment (Rheinberger, 2014). The descriptions by Robert K. Merton and Hans-Jörg Rheinberger refer to science. However, one must note that these construction processes of science, as they occur here, are equally applicable to the knowledge production processes surrounding innovations and technology. In both cases, one deals with the accumulation processes of knowledge and both areas can have an equal impact on society, both positively and negatively:

> [A] number of ideas that today we consider false actually changed the world (sometimes for the better, sometimes for the worse) and […], in the best instances, false beliefs and discoveries totally without credibility could then lead to the discovery of something true (or at least something we consider true today) (Eco et al., 1998:VII).

Thus, the question concerns not only the circumstances of an idea and its originality but also its assertiveness, which can vary greatly depending on the presenter (Merton, 1968). It is also interesting to look at the narrative of the concept in the context of innovations in their environment, e.g. how innovators and their surroundings operate with the term, as it has some predominant role for some time when searching websites or investors magazines. Observing serendipity in the innovation process can be disconcerting, especially when considering subsequent success, as it becomes challenging when success appears to be a matter of chance. However, investors and entrepreneurs insist on employing the 'principle of chance' to generate innovations.

This perspective progressively alters the demand for problem identification. If a potential inventor is not open to the chance events that can occur at any time, accusations of exclusivity may arise. The art of innovation, and indeed its demand, lies in maintaining openness despite the pursuit of problems. It is not necessarily the diversity of the solution to a problem that matters, but rather the *diversity of problems* to be discovered. In the initial stages, the solution takes a backseat.

This is where the race and pressure for the longed-for *eureka* moment begin. On the one hand, it is the search for the problems when they are not yet known or the race for a solution if at least one problem is already known and recognised. Alternatively, it comprises the drive to become known if one is the inventor of a problem and/or the potential solution. It is not uncommon for a solution to a problem to go unrecognised. As a result, society often overlooks its relevance (cf. invention of the electric car by Gustave Trouvé in the 1880s). In other cases, one can consider the solution to a recognised problem for which the original idea provider remains unrecognised, and someone with a better network takes the credit (Yaqub, 2018). Social structures such as networks, gender, money and the resulting competitiveness and resilience profoundly impact an idea's success, not only because they are the stabilising factors but because they can control the perspective on acceptance and benefit. No matter how good their ideas are, individuals categorically disadvantaged by the aforementioned structures often remain unseen. Thus, whether one can speak of chance discoveries—given the factors outlined above—remains open.

Serendipity is a phenomenon that has been sufficiently described in innovation research (e.g. Kingdon, 2013). What is interesting here, however, based on the theory described in Chapter III, is the following focus: if the emphasis is on discovering problems – partly because financing structures, like incubators, inadvertently necessitate this – then, as mentioned, the innovator finds the solution in the experimental space.

Furthermore, during the development of an idea, creativity might operate within a restrictive framework, specifically one that is problem-based (see subchapter 3.2.1.). For chance or the possibility of discovery in the sense of serendipity, this can mean restricting imaginative powers that cannot operate despite the required transparency in the problem-and-solution-finding process. Especially for innovation, the lack of openness can become a problem due to the simultaneously enforced closedness as, after all, innovation requires an open mind, united forces, and stimulation through the exchange of fruitful ideas. Furthermore, the joint decision-making process generates something like luck, which results in potential success (Elias et al., 2012).

However, assessing success and failure is independent of luck, even if it is considered a factor. In this respect, the question arises to what extent it is really a matter of chance or much more a matter of the plannability of something that some call chance. Inquirers must create real-world conditions, plannability, and experimentation in spaces that offer everything. Thus, random generators and algorithms provide a remedy for not relying on the randomness of chance. Predictability is more popular than pure serendipity when investing money and ultimately justifies the structures created around innovation. Nothing can be left to chance when too much depends on social factors. In this respect, serendipity is sometimes nothing more than a motive, perhaps a myth. The actual results of an evaluation are ultimately a mixture of chance factors, which, however, depend on the problem situation and not on chance as such, and probabilistic factors that try to make a statement concerning the future.

4.3.3 The Problematic Verifiability of Innovation

Despite all the constraints involved, serendipity and failure assessment practices seem to be highly prevalent in innovation-making. What is striking about these terms is the unpredictability that resonates in these construction processes, whereby the uncertainty provokes the temptation to give phenomena such as serendipity a special status. Similarly, there is an undeniably positive attitude towards failure. There is even talk that 'Innovation needs failure!' as in the *Museum of Failure* (MOX) slogan, which started its travelling exhibition about failed inventions from around the world in 2017. The exhibitors introduce their visitors to *The DeLorean* (1989–1990), *Logbar Ring* (2014–2015), and the *Boeing 737 Max* (2017–2019) – all of which are failed innovations. The MOX is not alone in its opinion concerning a productive view of failure (e.g. Wills, 2019), and it is common in innovator circles to approach work with precisely this attitude. How is this call to be understood? Beyond all doubt, there is a certain irony in this postulated acceptance of failure as, while it is accepted or even openly stated that there can be no success without failure, no innovator wants to fail. Failure does not lead to successful innovation, thus justifying the reinterpretation of failure as an act of mindfulness. What is necessary is not simply ignoring one's potential failure but rather adopting a 'mindful (approach) to failure' (e.g. Mielke, 2021: i.a. 26, 33), which must be preceded by an act of consciousness. Following this train of thought, it is said that that mere failure is insufficient for reflective handling of failure, as it does not allow one to recognise what they have failed at. The general reference here is how failure is managed, possibly through a corporate culture that permits failure and makes it tolerable. Consequently, the reflexive power of 'mindful failure' comes into play, which involves retracing the path and uncovering the sources of error – a formula for success based on previous failure.

The connection between innovation success or failure and emotional association occurs at different levels. For example, team dynamics and social structures, as described above, impact successful product completion. If the team structures are unstable, driven by conflicts or non-consensual goals, the idea can quickly be abandoned (process innovation). Furthermore, an innovation might fail if the assessment of user needs is inadequate or if the invention lacks general interest and thus does not have a market that could make the innovation appealing (product innovation). The further development of the previous figure illustrates these potential failures, which it had already suggested.

Figure 7: The Non-Linear Evaluation Interactions

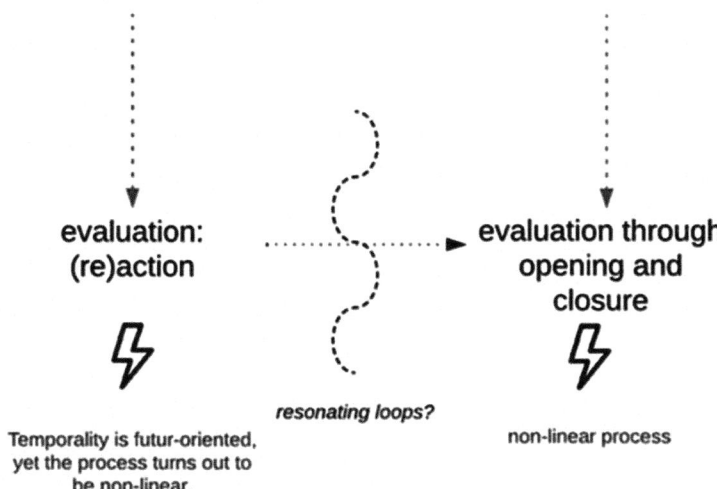

All these aspects can also be subsumed under the term problem-solving strategy to give the container concept of innovation some meaning. Exciting phenomena come to light in this context because one makes a virtue out of necessity. In recent years, in addition to so-called sprint sessions and *makeathons*, i.e. brief but intensive prototyping workshops that sometimes last a whole weekend, so-called *Fuckup Nights* have been established.

Interestingly, these Fuckup Nights are just as popular as *TED Talks*. Well-known innovators, investors, and (neo-liberal) politicians give speeches about the virtues involved in failure, the value of failure, and destigmatising failure. These events seem like an attempt to provide transparency to a complex black box system where there should be none, and they feature loose, generally positioned advice shared among entrepreneurs. The fact that an entrepreneur is willing to share their experience about their (prototypical) innovation with a random, broad audience does not seem surprising, as an invention in its infancy or an already failed project does not allow for any statement to be made about the development and, due to its vague formulation, ultimately also does not mean offence for the person giving the talk.

In addition to the positive portrayal of what no one wants to take away from the bitter note, the question remains whether other emotional aspects play a role alongside this to excite the failure. Is it a resilience technique or the old principle of 'trial and error'? These are coping strategies of the innovator scene and an optimising society that is stuck in its belief in progress and cannot put an end to a failed project. Failure cannot simply remain a failure and hence must follow a certain market logic of 'it was worth it' to potentially embed it into another linear storyline. Money that was invested but has not generated any added value in the sense of success is not justifiable and must, therefore, not appear in any narrative. Otherwise, as described before, no myth would be able to emerge.

Moreover, the irony is to be avoided after all, and one would prefer not to use a justification strategy. In that case, the difficulty remains concerning how to tell a good idea from a bad one (or, at least, a feasible idea from a non-feasible one). It is never foreseeable whether an innovation or what is called an innovation, will be successful. However, despite this, the question arises of whether it is possible to predict innovation success, as presented in many economic papers, by examining, for example, the error culture of a company. Are scenarios like the *Theranos* example mentioned at the beginning of this chapter avoidable? However, cases such as *Theranos* or *Amabrush* provide new evidence to open the discourse on the possible predictability of (mis)success (Ioannidis, 2015). As the innovation sections of magazines and newspapers have shown, they themselves are not reliable since they rely on assessments that are not objective or scientifically verifiable. Even so-called 'experts' cannot predict success and failure, which is solely due to the lack of transparency. 'Fake it, 'til you make it' is a solid motto that can keep an entrepreneur afloat for a prolonged period, even in very challenging phases. In part, this is an essential option for those who must overcome problematic development phases and hurdles without preceding their business partners' support. On the other hand, however, there is always the possibility of feigning success where there is none.

As Cristea, Cahan, and Ioannidis (Cristea et al., 2019) have already pointed out in their paper, the community must generally question to what extent the start-up scene conducts 'stealth research' because, as correctly indicated, fraud becomes public at a certain point, whereas unscientific work is not necessarily exposed. However, there is a well-founded desire within the sciences and society to share the scientific findings surrounding innovation. Other ethical measures apply mainly in the medical field. While it is not necessarily of collective interest whether the next 'smart device' will be successful, entrepreneurs become accountable (at the latest) when they claim 'social impact' for themselves and their product(s). This impact arises when technological inventions are disruptive to such an extent that they influence legal, ethical, and social issues and can potentially cause harm, which is the case with Theranos or potentially with automated technologies such as smart cars. This question boils down to what criteria societies use to evaluate an idea and, consequently, a prototype and to what extent these criteria are linked to the abovementioned expectations or to what extent they are emotional, i.e. subjective. A simple example of this encapsulates the question concerning to what extent the desire for a solution is more dominant than its actual reliability.

Chapters V, VI and VII present and discuss the case studies and, in doing so, refer to the theory presented in III and IV. In the current examples, we examine the genesis of an idea in its initial state, addressing both the emotional connotation and the subsequent modifications that the notion and its related players undergo as they evolve. The notion prompts the evolution of the moral economy.

The investigation begins with the envisioned concept serving as a projection surface for dreams and futures in response to one's difficulties. It scrutinises the challenges and obstacles to growth and the techniques for surmounting them. In conclusion, we examine the assessment throughout the concept's development, its materialisation in the prototype, and its outreach and (re)claiming.

Empiricism – A System of Emotional Forces Around Innovation

Innovation is a term often associated with new technologies, advances, and practices, as shown in the preceding chapters of this book. The phenomenon is also challenging to examine. As evident in Part I, Chapter IV, the issue stems from the incomprehensibility of the term's inflationary meaning and deliberate opaqueness. When one attempts to maintain a careful check on prototypers, inventors, and innovators, it soon becomes evident that such undertakings are often not permitted. This situation is particularly challenging for ethnological research such as the study undertaken here, and there are barriers between ethical openness and inventive reluctance relating to this work that sometimes appear insurmountable. Some of these issues may periodically reappear in the following descriptions because they illustrate this study topic and thus also serve to contextualise the research findings.

Despite the challenges outlined above, this chapter provides insights precisely into the everyday practices of innovating. It is thus not a matter of stringing together case studies in their respective environments but rather conspicuous features that have a congruent character for the examples, meaning that the following subchapters do not deal with one prototype example at a time but with phenomena that occur independently of the prototype, the team, or the space of development. Consequently, I will not present the aforementioned teams and individuals, their ideas, and their development environment sequentially. In the subsequent subchapters, I will introduce all the study examples and refer to phenomena that ethnographic research has adequately documented, providing information about conspicuousness. It is worth noting at this juncture that studying prototypical artefacts and stages serves as a means to an end in the study of innovation development. The prototype offers material advantages as a mirror of innovation practices that act as an example. As described previously, the prototype is a testimony to processes that are otherwise poorly documented. In this respect, examining a prototype offers support in the investigation alongside immaterial data whereby the concept of a prototype can be broadly defined and sometimes has to be.

The claim and the potential of innovating, whether implementing the idea in a tinker's cellar or a large company, always lies in a changing measure to achieve success. It initially involves the idea of a cure, preferably for a specific problem. The everyday problem thrives to become the breeding ground for all solutions that are supposed to change something positively, and it is precisely this phenomenon of world-changing action that is omnipresent in the available data. It is not the desire to be creative alone that drives people to build something. In a 'what-if scenario' that unfolds in the future, the imaginative remedy emerges from the projection surface of everyday life. The emotional waymarks of imagining are included and presented in the following examples. Apart from the aspiration to change the world, the tirelessness of this aspiration holds dangers, an enormous feeling of ecstasy and the belief that carries one. Therefore, in the context of the information provided in Chapters

III and IV, the following four subchapters discuss the four stages from having an idea to actually convincing others of it.

V. The Imaginative Remedy

5.1 Dreaming of Bright Futures

> **From the field diary 6 February 2020**
>
> *I am looking for a prototyping lab on the edge of the Berliner Ring, just behind a large park within the city. I discovered this lab on the internet and found the name appealing and seemingly meaningful, so I subscribed to the newsletter. Now, a few days per week, I receive around five e-mails inviting me to laser cutter workshops, wood workshops, or 3D printer workshops and so-called 'meetups' where already successful entrepreneurs explain how I can 'realise [my] dream of [my] first product' (e-mail invite via the e-mail distribution list). Sounds ambitious to me.*
> *I walk along a street with container buildings—the residential area ends here—and I try to find the lab. Everything is relatively inconspicuous, with no descriptions, no signs to show the way and no house numbers. I enter a fenced area, walk towards the first container hall, and find a small sign: 'M.lab'.*
> *Here, I will interview one of the lab's founders. I enter the hall, and the atmosphere changes abruptly. It is the soundscape of a school gymnasium. People run around, greet each other, quickly ask each other how things are going, and disappear behind retracted walls and grids—their workspaces, as I would later learn.*
> *All kinds of equipment are standing around, and the place is chaotic and untidy: high ceilings, tables, old sofas, and many tools. I stand in front of a bar and ask for my interview partner. [...]*

This excerpt from the diary reflects one of my first impressions of a prototyping lab. It is not an incubator but a form of collectively managed space that one can rent monthly to realise one's idea materially. Individuals and teams can find a space and use tools according to their abilities and needs, such as the laser cutter described above. The monthly rent covers the use of the tools, with the sole security measure being the necessity to attend pertinent workshops to enable their use. The lab I visited housed two large container halls during the interview period. The operators partitioned one of these halls into plots for storing and distributing products of pre-existing start-ups. The other hall houses several teams that are still working on their prototypes. At the entrance, there is a bar and a communication area for exchanging ideas, planning, and eating, and next to it are other plots. On the one hand, there are 'offices', which some call their plot, and on the other, various tools such as 3D printers and workbenches. There are further levels, which I reach via industrial metal stairs. The hall offers more space than one would expect from the outside. It seems labyrinthine, and I repeatedly discover niches where people are working on their projects. This place is a place of fulfilment for some, while for others, this is where the realisation of their dreams begins. The sentence that jumps out at me in one of the invitation emails to the *Inventor's Night*, 'How you realise your dreams', comes to life here for one or the other person. Through casual conversations with the inventors in this lab, I gained insights into the ideas they plan to implement here. I meet sex toy creators as well as inventors of new bicycle drive models. What unites them is the full conviction of the value of their idea and the belief that they are pursuing a greater purpose.

In conversation with Christian, one of the founders of M.lab, it becomes clear how the idea of something better and bigger manifests itself in the objects:

What would an inventor or innovator be without his idea? After all, the idea is what innovation is all about. An [can be a dream of a better world]; it somehow carries the character of utopia. In general, the idea always refers to a "could-be". *(Interview from 06/02/2020, Christian, Founder of M.lab, own translation of the German transcript)*

The conversation with Christian clearly shows that the idea is part of the inventor, that it defines them. They speak for each other situationally; they mirror each other's experiences and (emotional) values and pictures of *a* future. Further, in each phase of the idea's development, these imaginations are a projection screen for inner and outer negotiations. As subchapter 4.1. described such makerspaces as giving space to the countless hypotheses someone has developed in his imagination – a 'could-be'. These visions of the future may be dreams of a better world, as Christian says, utopias and potentials of possibility. At best, the idea and the potential result should improve the world and change it for the 'better', although what precisely 'better' means is not further defined. However, it depends on temporary collective value definitions, and the reference to an improved state of the world is by no means exaggerated. The fact that, in addition to the state of possibility, reference is made here to the enhancement of utopia makes it clear how inherent this desire is in the description. Consequently, it can certainly be more than the mere and supposed improvement of something already existing.

Figure 8: Inside the M.lab – Container One aka Utopia

Utopia as a 'non-place' and counter-design to the status quo thus describes what does not yet exist. *Utopia is the prototype as an artefact that has not yet been realised.* Space and things become blurred; the prototype becomes the expression of a utopia; as a part, it originates from the vision of the non-place. The idea space 'topos uranios' (see subchapter 4.1.3) is the heavenly place where the idea dwells. In this way, the place of innovation can equally be perceived as a utopia, whereby the prototype is a parvenu of this place, which can only come about through one's imagination. The incubator or the makerspace thus does not yet make a utopia: it needs people and their imagination to make this place their own and try to bring their ideas to life and, consequently, to enliven the space. This form of resuscitation requires a commitment that arises from intrinsic motivation. Where this motivation comes from ultimately varies from individual to individual, as will be shown in subsequent sections. However, one aspect is already visible: emotions, fears, insecurities, dissatisfaction or anger, and sheer passion or inflamed conviction often accompany motivation. All these accompanying emotions say something about our relationship with the world in which we live. They represent our attitude to what we have experienced, and the ideas we develop are ultimately a reaction to them.

This assessment also aligns with Karwen's statement. As an innovator and private investor, he does not immediately speak of imagination but notes that it is indispensable for the idea's process.

> To be honest, I've never thought about imagination but rather creativity. But well, when I'm thinking about it now, I guess without imagining something, it's just or probably not possible. So, it's just not possible that I want to invent something and don't think about it beforehand. Something always happens beforehand somehow, then I think about it, and then I usually know what I want to do. Or, also in another way: sometimes, in a conversation or something, I suddenly think, wow, that doesn't exist yet. We have to do that! It's a sudden idea that just comes to me. Of course, I didn't think about it for so long, but the idea is still there. It's also fun, somehow. *(Interview from 18/06/2021, Karwen, Private Investor & Innovator, own translation of the German transcript)*

Karwen describes the process between the inside and the outside world, the oscillating moment of perception of the outside that becomes *an* experience, from which one, in turn, nurses one's ideas and thoughts (see subchapter 3.1.1). Franz Brentano calls these psychological phenomena the outside world that is sensually perceived, and if one does it consciously, John Dewey would speak of *an* experience. From his point of view, Karwen mentions that something that triggers his imagination first has to happen. Hence, it takes conscious perception to evoke something in himself – an *object of inner perception*. The conscious act of perception arouses the act of creativity (see subchapter 3.1.2). In doing so, he describes the idea in two ways.

On the one hand, something takes time, progresses further and goes through a conscious process. The creative act may stem from new *additions* to what already exists or from new *combinations* (see subchapter 4.1). The ultimate assessment of these constructions as innovative hinges on the evaluating actors' ability to compromise and the acceptance behaviour of a society. However, as shown later, much of these evaluation practices depend on a culture and a narrative that emerges from it that contributes to the character of a product.

On the other hand, something emerges suddenly as an idea and becomes manifest, which Byrne calls *insight* (Byrne, 2005: 193 f.). This is a spontaneous idea that arises at the moment. Byrne would say there might be no reflection beforehand; however, the creative moment remains a reaction to my surroundings. Later in the conversation, Karwen clarifies that ideas take time to develop. Whether the sparking idea suddenly appears or only solidifies over some time says nothing about the subsequent development process.

Either way, it ought to solve something—either a problem or a lack of something. Consequently, it means a disruption, the moment of 'creative destruction' (Schumpeter, 1942) that develops something that renews what previously existed. In our interview, Christian elaborates on how individuals in the lab manage their ideas and the origin of their desire for disruption.

C: Mhm, you are always confronted with something. With us, it's usually like this: people come here and already have a mental picture [Vorstellung] of what they want to do. Then, they need tools. [...]
I: Let's go back to their mental picture. Can you describe in more detail what you mean by that? Are these already finished ideas, or how do I best envision them?
C: Yes, not necessarily finished ideas, I think. From what I hear, they have concepts about what their thing should do or be able to do. That's clear. You know, that "Swiss army knife" ["eierlegende Wollmilchsau"] thing. Something super great, perfect. *(Interview from 06/02/2020, Christian, Founder of M.lab, own translation of the German transcript)*

He indicates that the desire for change is already inherent in the imagination, more so for disruption, in terms of the 'Swiss army knife' he mentions. As Christian says, the idea is an all-rounder, at least in the imagination, able to fulfil every exposed need. In the vision, obstacles do not arise, or if they do, only very incidentally and not as an insurmountable problem, along the lines of 'I have identified a problem or a shortcoming and have the solution up my sleeve in my imagination'. Hence, utopia lies in my imagination. In this place, the idea is already an independent entity that develops a metaphysical reality. In this reality, the idea is present as an entity that also acquires agency, but it equally contains my values (see subchapters 3.1.1. and 3.1.2.). Through the ability to direct feelings towards something in my imagination or to revive emotions that I have felt through a lived experience, I can create a connection between myself and the outside world. More than that, I can even review my feelings from the past and reconcile them with my values in the present. The moment I decide how to solve a problem through an idea is already the situational judging moment of a problem or a disproportion. I consciously experience and feel, and after that, I judge and am motivated to make something different out of it.

It can be observed that for Karwen and Christian, as well as for the innovators in the incubator, the underlying canon of values inherent in the idea forms the basis for an initial narrative of creation, which comprises arguments for implementing an idea that a larger group supports whose needs or even just wants are addressed. I will follow up on this later in Chapter VI.

Justification patterns for inventive purposes emerge from this value complex. The purpose is the solution to an experienced, observed problem in everyday life, and it is justified by its curing effect.

5.2 Seeing Problems, Being Radical

As previously described in theory, the data material shows a clear tendency for problems that are perceived in everyday life to give rise to a notion of a 'problem'. In the following section, various interview excerpts show that creativity or the will to develop something often stems from observing everyday problems. As previously indicated, one can hardly separate the imaginative act from experience. They mutually reassure and condition each other, shaping the emotional attitude towards oneself, the external world, and one's actions. The pragmatistic triad manifests as thinking, acting, and feeling. Thinking corresponds to what we previously described as conscious perception. In this context, acting transforms into a creative act based on my experiences and ideas. Finally, feeling integrates with the other two principles. I cannot think or act without feeling. Experience morphs into a reality from which my attitude and actions are derived. I *root* for the problem; I become – in the best sense – *radical*.

In a conversation with a consultant from the incubator, who assists the teams in planning milestones and provides advice on their daily tasks, it quickly becomes apparent that:

> This question, "Where does the idea come from?" is already being discussed. Usually, from the recognition of a professional – from a professional recognition. From everyday life. Especially now with clinical researchers. As a rule, they see the potential for improvement in their everyday work. It is rarely something artificial. So, someone says, "I sat down and did a brainstorming session, a design thinking process, and then I came up with this and that great thing." That's not really the case. Rather, people come from everyday working life. That's why it's so interesting to work with clinical researchers. They have their everyday lives. They see patients. And from that, yes, they do. So, that's the overwhelming number – if not all. I can't give you numbers. But it is clearly the majority. Improvements recognised from everyday life. Often, the scope of an idea is not yet clear. That's not an issue, either. First of all, it's about concretely improving something. This can be big or small.
> (Interview from 13/07/2020, Felix, Consultant at Health Hub, own translation of the German transcript)

The physicians who develop their ideas in this incubator have previously applied for this accelerator programme with an existing solution to a problem. It works in the following way. Firstly, the medical doctors who work in the clinic discover an unfavourable circumstance, problem, or deficiency. Medical practitioners can formulate this discovery in an application at any time and must include a concrete solution in the proposed solution section of the application. By doing so, they apply for a multi-year funding period that enables them to (partially) step away from the daily clinic routine and conduct research. If the proposed idea is logical and adds value to the incubator, the program accepts the doctor. After a further process, a team is put together capable of practically implementing the idea and making it usable. Later, the product should be sellable. In recent years, several institutions and even university hospitals have established similar experimental rooms and makerspaces in complete accordance with the objectives outlined by the BMWi at the time (subchapter 4.1.1.).

Therefore, becoming radical is to be understood in an exclusively positive way. These experimental spaces are created to scrutinise problems that are already known or to track them down. The protected space can help solve a problem by isolating a circumstance. The protected space, the 'lab' or the incubator cannot provide a simulation of everyday life. Everyday life is outside, in the world that I actively experience.

In the following quote, Bahar explains concerning her project that the biggest problem in her everyday life in the clinic was to remind the patients after a knee or hip operation that they should not put more than 15 kilos of weight on their affected side for several weeks to avoid consequential damage. Today, she and her team are developing an insole with sensors that measure the weight applied on each leg.

> The basic idea was always that our patients had to bear a partial load of 15 kilograms on their feet after the operation, i.e. they were not allowed to carry more than 15 kilograms, and actually, they didn't know what they were doing. They are allowed to do that for six weeks, but of course, nobody knows what 15 kilograms is. The physiotherapist says that once and then after six weeks, he [she] says "goodbye and have fun". And then sometimes they come back with nasty complications and don't know what the problem is. And these digital soles were meant to control this pressure management. The problem is that the soles were made for athletes at the time, and the studies that were running were always with super digital people in their mid-20s. When I somehow collected the last digital sole that I had borrowed from Brandenburg because the patient had dropped out of the study again, I thought, well, it can't be that these are the only soles that exist. And then, I did some research and found out that they can't measure weight. They only measure pressure. They can't say what five kilograms are; they say it's three minutes somehow. *(Interview from 30/01/2020, Bahar, Physician & Innovator at Health Hub, own translation of the German transcript)*

In this excerpt, she describes the predecessor models that could not solve the observed problem. They were insoles that could measure pressure but not weight and were not practical, especially for older people. The problem and the lack of a solution made her reflect, and finally, she tried to develop an insole herself. It was elementary at first, a preliminary prototype that her sister welded together with a few components in the basement before they both went to the accelerator programme of the incubator. At this point, it also becomes evident that Bahar developed this idea using a different approach than what Karwen described for some ideas. In her case, it was not a spontaneous idea but a more protracted process of (re)consideration that led to a potential solution to the perceived problem. This discovery approach also confirms Felix's observation that most of the ideas of those who apply to the incubator originate from their everyday work.

Figure 9: Sketch by Hendrik explaining the Prototype's Genesis

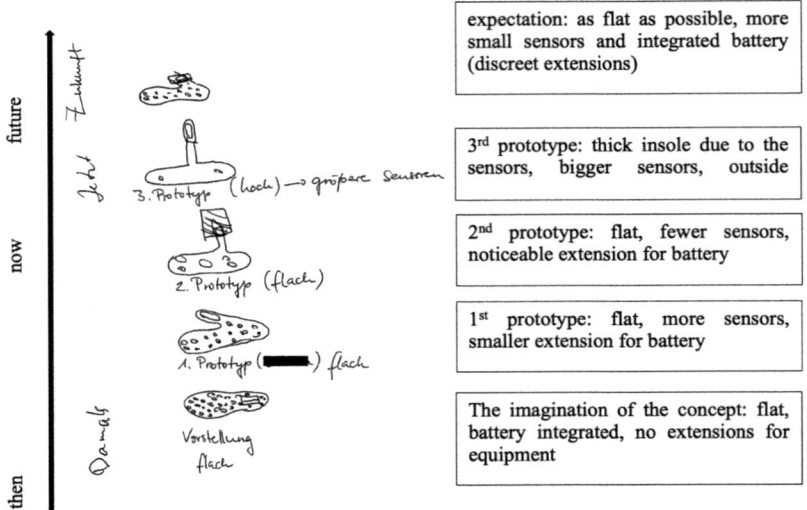

Figure 9 shows the prototypical course of the insole for hip or knee surgery patients with post-operative restrictions. With the first conception (at the bottom) and the first prototype, Bahar, a medical doctor, and her sister applied to the incubator in 2018. The bottom drawing shows the first presentation before there was a prototype. The subsequent pictures show the first, second and third prototypes and Hendrik's vision of the future product (top). When I asked Hendrik to make a sketch showing the prototypes, he did not embed it into the otherwise existing narrative, showing the mixed nature of the work steps taken until then. He took the direct route and only drew what was present as a materialised form.

Viktor is the product developer of this project and joined the team in 2019. He worked on a similar idea on his own before he joined the team. To realise his idea, he joined the incubator and the team to have a funding opportunity. In the interview, he tells me how he came up with the idea of developing an insole with sensors.

[...] In 2018, I moved from Singapore to Berlin with a program called Entrepreneur First. It was an accelerator. [...] It didn't work out for me. It was a very interesting program. You brainstorm, you get a couple of thousands of euros just to brainstorm for three months with 50 people, and you do pairs. So, very interesting process. And then, that didn't work out. I was still in Berlin. I started working on my idea. I had a couple of ideas like all hardware related, you know, one with bone conduction, another thing with an insole and the insole stuck with me. The way that this project started, in my mind, was in a hospital in Singapore. There were frailty patients who would not get out of bed, and we were there with the type of system meant to rehabilitate them [...] after frailty. Frailty is a disease that has different factors combined, and the moment that you get there, it's a combination, let's say, of muscle fatigue or muscle weakness and depression and other things. *(Interview from 04/02/2020, Viktor, Developer at Health Hub)*

Figure 10: Third Prototype of the Insole

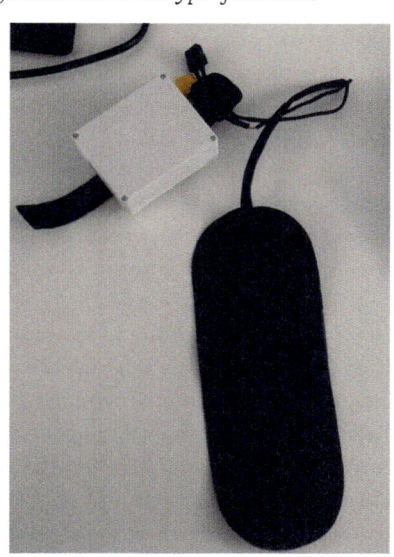

Viktor is an entrepreneur, innovator, and product developer, not a physician. He was asked to develop another product for a hospital through his previous work at MIT in Singapore. Through his hospital experience there, he was aware of specific patient ailments and problems and later drew on them in his work to look for solutions.

As Viktor describes the approach, being an innovator is about collecting problems and finding the potential for bettering the world. To achieve this, one must thoroughly understand a problem's everyday complexity. This evaluation implies that a problem is not a static phenomenon that consistently manifests similarly. It changes with its actors or the bodies of its actors, with the different ways of dealing with it, and with the times of the day. A problem is never just 'one'; it has several facets and is rarely uniform. However, for both Bahar and Viktor, it is the starting point for their solution and is also highly emotional. In the subsequent sections of this chapter, I will show how problem perceptions are linked to emotions.

5.3 Emotional Motives

Experiencing a problem emotionalises and can be the seedbed for the creative moment whereby these moments from which a motive arises are individual, and socialisation and everyday life shape the motive for an idea (subchapter 3.3.1). Emotional motives occur in the individual frame of reference or a thought collective (Ger. *Denkkollektiv*). Motives and emotions are mutually dependent since a problem is evaluated individually within the framework of what has been socialised and learned. Thus, the problem evokes a feeling in the person concerned. Simultaneously, the innovator responds to the problem and the evoked feeling with a solution from the same frame of reference. These answers reflected in the prototype are the materialisation of their reference system of feelings.

In their personal accounts, the interviewees vividly express their *frustrations* with the shortages. They recount instances where they have had to grapple with poor or missing problem solutions for diseases they encounter in their daily lives. These shortages, be it in equipment, staff, or time, often intertwine, causing a significant impact on their work or a sense of *compassion* and *burden*. Their motivations, they reveal, oscillate between idealism and the desire to create something valuable and marketable. In a figurative sense, as per Arlie Hochschild's theory, inventors' solutions are grounded in their respective referential systems; their proposed solutions are part of a controlled system fitting into a manageable society without intrusive (Hochschild, 2012: 4). Furthermore, these solutions only reveal the aspects that the inventors are prepared to disclose. This constraint through control is the nature of emotions and the innovations they inspire – both have their limits.

Bahar, the doctor at the incubator, shares a poignant *experience* that underscores the emotional consequences of shortages. She recounts a situation where a medical need for sensory soles was not met. As a result, complications arose after surgeries as patients did not fully comply with the recommendations. This situation, she admits, filled her with *anger*, which became her driving force in developing a product that could address this grievance.

> **I:** Where does your motivation come from?
> **B:** To do that?
> **I:** Yes. Or where did the original one come from? How did it come about, this post-op?
> **B:** It's always a bit stupid to admit it like that. But I was incredibly angry that there were no soles that I could use. [...] I was very angry, and I couldn't afford other soles

either because I didn't have any funding yet. And for me, anger is a good motivator.
(Interview from 30/01/2020, Bahar, Physician & Innovator at Health Hub, own translation of the German transcript)

Interestingly, she remarked that it is 'stupid' to express her emotion, i.e. her anger. When asked later why she thought it was stupid, she replied that it sounded pathetic and pathetic sounded trite in the context of her profession. The passage can only exemplify that the expression of emotions in connection with scientific work has a tense relationship (subchapter 3.3.). However, the emotional context is indispensable for expressing the motive. In this case, it even provides credibility. Bahar sees herself as an assertive young doctor who feels an inner drive. She describes her anger as a strong motivator. According to this, there is a direct relationship between her feelings and the work or the motive for her idea.

Figure 11: Trying Out the Insole in an Orthopaedic Shoe

Contrary to the common expectation for scientists to depersonalise their work, Bahar's display of emotion challenges this norm (Daston & Galison, 2007: e.g. 52). This contradiction suggests a departure from the traditional scientific practice of maintaining emotional neutrality, highlighting the potential for personal and emotional engagement in scientific endeavours.

It emphasises that scientists like Bahar are not just detached individuals but also have their own emotional responses and personal perspectives. This 'personalisation' underlines the emotional dimension of scientific research and innovation, making it a more human-centric and relatable process. This interview with Bahar is more than just a language exercise. The journey reveals the transformative power of personal context and emotions in scientific discourse. As Bahar shares her scientific work, she does not limit herself to objective reporting. Instead, she weaves her everyday experiences and the emotions they evoke into her narrative. This linguistic turn in her storytelling changes the content and choice of words and alters her thinking and feeling in relation to the prototype. It is a testament to the profound influence of personal context and emotions on scientific discourse and research.

Bahar's product developer, Viktor, initially also had a similar idea that he wanted to implement independently of the project. He ultimately points to several motives and emotional connections. First, he describes a feeling of *confinement* related to his experience in the Singaporean hospital.

In this interview excerpt from the 15th of January 2021, he solely refers to his medical product ideas and motivation. The feeling of constriction he describes indicates a sense of *compassion*, a sense of *empathy* that he feels for a problem situation, that is, the ability to put himself in a situation and here, above all, the reference to imaginative power becomes evident as it takes imagination for one to be able to put oneself in an (emotional) situation (e.g. Barbalet, 2005: 178; 2006: 51; Villanueva, 2012: 139). Furthermore, Viktor shows what matters to him and in what he is willing to invest feelings. In subchapter 6.1, we will discuss the incubator as a place where individuals exchange motives, ideas and feelings that they are willing to show and negotiate (Hochschild, 2012). The incubator is a space where individuals can share their emotions and thoughts with others, and these exchanges can lead to the development of new ideas and perspectives. By actively engaging in these conversations, individuals can gain a deeper understanding of themselves and others, ultimately leading to personal and professional growth 6.1. The interview with Viktor highlights that the purpose plays an important role. He repeatedly mentions purpose and impact in the interview, which will be addressed in more detail later. At this point, it becomes evident that he opens up a personal reference, which again refers to a highly emotional situation. He thinks of his grandmother and refers to the walking behaviour of older people, which he perceives as everyday difficulties:

I: Did you feel some kind of compassion, or why did you think you could or wanted to develop a solution for this?
V: Yes, probably. There were a lot of problems, a lot of things were connected, and you have a feeling of confinement. […] So, then, from healthcare, my motivation was to help members of society who have difficulties, let's say, visually impaired was the time when I switched to healthcare. And I said, "Okay, that would be really

useful. And it would make more sense for me to do that." And then I switched to [the] elderly because, for example, one reason was my grandma. You know, I noticed that people who are older have a different way of walking, and I thought this would be really easy to understand and to monitor and detect differences in the way of walking [...]. *(Interview from 15/01/2021, Viktor, Developer at Health Hub)*

It is clear from both the conversations with Bahar and Viktor that a *personal connection signifies* and points to an (at least initial) idealism whereby *compassion* plays a role in both. Through this compassion, they are connected to the patients on an emotional level and draw their motivation from this. In this context, compassion becomes a compass or navigator and has a guiding effect. It provides an ideational destination route.

In the following excerpt from the interview with Ryan, who is also working in the incubator but on a different project, Ellie, the motive is self-referential. His emotional motive differs from the other two in that it is not about a personal relationship or an idealistic attitude towards an observed deficit. The original motivation for his current activity refers to him and his memory from his childhood. However, it is precisely due to the memory and the resulting nostalgia in his narrative that the motive is no less emotional. In this excerpt, I ask him why he followed his supervisor's idea to apply to the incubator *Health Hub* and what made him so enthusiastic about it:

I: Okay. So, how did you actually suggest taking his idea to the incubator? [...] What was your intention, or your motivation, rather?
R: Well, the intention goes back quite some time because, as a kid or basically going through medical school, I always wanted to build something. As a doctor, you're a user, which means people and companies build these technologies and products, and you, as a doctor, you use them without really any thought of how they were built. What do they actually measure? How are these systems updated? What's the technology in it? I just get it, I use it. I have to trust, okay? A value comes out. And I just, I have to just know, and this, for the most part, 95% of physicians, imagine they're happy with this. I'm just in the small cabinet of physicians and say, I actually want to build the thing. I don't want to just be a user. And as a kid, I always wanted to build robots. I was always interested in building computers and robots. I never went to informatics, unfortunately. Kid of the eighties, my parents didn't see much value in computers, which, again, I'd recommend this for a kid now, get them learning right away. So, I entered medical school. I learned everything. I was much more interested in science with it. And just, I guess it's just a coincidence. I happened to be at this one anaesthesiology Department with a boss who had this idea that needed someone to actually use it. And the benefits where I can finally get out of the clinic, I can actually build something. If it works, then there's an instrument in every single operating room. And I can say, okay, I had a

> hand building this instead of just purely just using the stuff. *(Interview from 15/10/2021, Ryan, Physician & Innovator at Health Hub)*

First, he talks about a childhood experience. His memory, a source of emotive forces (see 3.1.1), reflects an experience that he, in turn, links to a will to create. The act of creativity is emotional for him because it triggers a feeling of *excitement* in him.

Above all, Ryan is motivated by the idea of creating something by himself and not just relying on others' technology. He talks about 'trust' and 'value' in this context and opens up to perspectives that seem relevant to him: the user and the builder. Ryan compares the two persons and provides them with attributes of passivity and activity. He thinks the doctor is ultimately just using the technology provided to him without thinking about how it works and what it does. Consequently, the user must trust the technology's functioning to generate value on which he builds or justifies his further work. Thus, Ryan counts himself among a minority of doctors who oppose this. He considers himself part of an active minority and sees this as founded in his childhood because he 'always wanted to build robots […] and computers'. He now sees an opportunity in the development of the device. He considers this satisfying his interests, which is why he feels so much *enthusiasm*; he sees the work as *fulfilment* and himself as a medical researcher instead of a practician. If it works and the idea becomes an implementable success, he will have created a value according to his moral concept. Value projections of this nature are present in every operating theatre.

The feeling of enthusiasm can also be observed in Karwen's explanation, although he justifies it differently. Admittedly, he also speaks of a purpose to be fulfilled and mentions idealism, although the latter aspect seems to have a half-life. He told me that if a project seems hopeless, he will not develop it further or invest effort in it. Idealism does exist, but more in the uniqueness of an idea than in providing a solution for a given need. For Karwen, it is about demands, not necessities, and the attraction lies in developing something that does not yet exist. Additionally, it is about marketability. Karwen's ideas do not 'just' involve inventing something; his approach is practical, and he has to be able to make a living from it.

> **I:** What is ultimately the motivation for your development or ideas?
> **K:** That depends.
> **I:** On what?
> **K:** (Laughs) Look, most ideas are supposed to have a purpose. Of course, it can be idealistic, maybe even most of the time. (I: Mhm.) And then you also want to sell the things, of course. You don't make the effort if you don't hope to get anything out of it. Inventing things just for the sake of it is not so much fun.
> **I:** So, what, does the idealism disappear over time then?
> **K:** Yes, I don't know…let's say that the idealism of being able to live from it is higher, isn't it?

I: OK. Let's go back again. When you think about your last project, what originally fascinated you about it then?
K: That it doesn't exist yet. It's the most beautiful thing to do, something that no one has thought of yet. That's the best, you know. This really—I don't know—I really love that kind of feeling. It's a—yes, it gives me power.
I: Oh, that's a lot of emotion there. What is the ideal scenario? Is it the one where you're able to invent something that doesn't exist, and you're able to sell it?
K: Yes, defo.
I: And if it's not sellable, you won't go further with it?
K: Ehm, then you might ditch it sooner. Sure, it's just a matter of eliminating some deficiencies. You build a platform because it should help people, it should make everyday life easier, and should connect people who benefit from each other.
(Interview from 18/06/2021, Karwen, Private Investor and Innovator)

Karwen's financial success is evident in his ability to invest in other start-ups as a private investor, often in the five-digit range, when convinced of the project's potential. During our interview, Karwen often mentions the factor '*fun*'. Something should be fun, for example, inventing an idea, as he describes in the previous excerpt. Innovating as an end in itself, on the other hand, is not what he prefers. He talks about purpose, but I get the impression that, to him, purpose purely refers to marketability. However, this does not diminish the feeling of *excitement*; it just has a different impetus. If one tries to define the word excitement, one encounters descriptions such as: 'an endeavour that gives pleasure' or an 'activity or situation that makes one feel happy'. Accordingly, Karwen feels *joy* when he perceives himself as a 'first discoverer' and successfully convinces others of his invention. 'It gives him *power*', he says. His creativity, actually discovery, takes possession of him, which is a feeling he *loves*.

Whether Bahar, Viktor, Ryan, or Karwen all show their motives for taking on a problem and finding a solution are emotional. Even if the motives are different in that they do not seem exclusively altruistic, they remain emotional in their description. The same applies to the discovery of the problem. Even if Ryan did not encounter the problem himself, he found access to it through his supervisor and took it on, whereby the emotionality nevertheless remains. In his case, it is not only his own discovery that makes the motif emotional but also the fact that he can identify with it or develop empathy, that is, create a state in him that develops an awareness of the problem and thus opens up space for an emotional reaction.

5.4 Conviction, Purpose, and Impact

As has become apparent in the interviews, purpose plays a prevailing role for the interviewees. Overall, the interviews make clear that belief in a 'good' idea is a prerequisite for innovation. In this context, 'good' refers to the understanding of the respective speakers, and they have a relatively concrete idea of it.

> So, I had this start-up experience. And then, after my PhD, I needed to go in one direction, you know, to do something else. It's called, let's say, post-PhD depression, you know, where you want to... you achieve something, it's ready, it's beautiful, your creation, but now you have to find another goal. So, after that, the postdoc was a kind of transition. And after that, I wanted to do something more meaningful, let's say, with more impact. [...] back then, when I did the Facebook app, it certainly had a huge impact in my country, because I was going on the street and everybody knew the app, and I was just doing random sampling. Everybody knew it, you know. So, that's a very high impact. However, let's say my opinion has changed. And even at that time, I had some problems with my conscience, the fact that it's not making people very [much] smarter or better. So, at that time, in 2009, I had these types of issues. And yes, what I'm doing right now, I believe, it's a different level of impact. It's high. It's difficult to quantify exactly the impact, first of all. But this is combining, right now, different types of interests of mine. *(Interview from 15/01/2021, Viktor, Developer at Health Hub)*

In one of our conversations, Viktor tells me how much he has been influenced by his past developments, such as a social media app he developed in his home country in the early 2000s. He is also aware of the high level of impact, but after his PhD, he realised that a *different kind of impact* is more important to him. He speaks of 'huge impact' and 'high impact' but is aware that what he, in fact, describes is not possible to measure but rather refers to a felt impact of what he considers worth doing. He thus develops a moral component for himself as a means to a 'good' idea. He summarises the different forms of influence under impact but now attaches importance to something I propose calling *moral impact*. He criticises his social media app for not making anyone 'smarter' or 'better'. Nowadays, he talks about high impact, focusing more on his interests. As he explained, he is concerned with improving people's lives with medical problems related to their walking ability.

Susan, the founder of The Believer School's creative space, also tells me how much meaning must be contained in work to develop something valuable for society. She talks about how she does her work on a more abstract, general level:

> [...] People don't explore inside them; they explore outside them. And I think that part of, that's the most important part of what I do. In fact, it's like, I really want people to love themselves, to think about who they are, all they've been through, all they've survived, and to be hopeful about what they can contribute to the world. And I use technology, you know, (?) chain learning, and computer vision VR [Virtual Reality] to get them to reflect on those things. Because otherwise, I wouldn't do it. So, that's kind of, I don't know, how I'm approaching this stuff. [...] I think it's

important for people to create something that comes from things they care about because also it's difficult to finish things, and people start things all the time that they don't finish. And I think that if you have a purpose and a reason for finishing, because it's something you care about and you want to explore, then you're more likely to finish. *(Interview from 12/08/2020, Susan, Innovator & Founder of The Believer School)*

Figure 12: Exploration Phase One at the Creative Space

When Susan leads projects, she must initiate a process of self-reflection among individuals, guiding them towards understanding their contributions, motivations, and passions and illuminating the path forward. In addition, she frequently emphasises social and moral components during the interview. However, there remains ambiguity regarding what individuals prioritise and value. It can be the question of my own emotional life, in the sense that I ask myself what is important to me, or also the question of what I wish for society and whether I can do something to support it. As with Bruno Latour (e.g. Latour, 2004), Maria Puig de la Bellacasa (Puig de la Bellacasa, 2017), or the ground-breaking work of Joan Tronto (Tronto, 1993), the subsequent taking care of my idea is not left out. However, this gives rise to several interpretations of care. Even if Tronto thinks that care has no self-reference (Tronto, 1993: 102), I would like to express a slight doubt concerning this since, even if there is self-reference, caring for others can still be encompassed within it. For example, a participant in Susan's workshop may well be interested in a solution to a problem of his or her own and want to work towards it if, in addition, the need for others is equally present. The phrase 'I care' is the phrase of multiple commitments expressed in an idea, problem-solving, and general participation in social life. In addition, it is also the responsibility for an idea that I am willing to take on beyond its 'birth into the world'. Even an invention needs further care later, especially when introduced to others. In the words of Annemarie Mol:

> It [the logic of care] assumes that things are just as unpredictable as people. It does not take technologies to be "mere" instruments. Instead, good care involves a persistent attempt to tame technologies that are just as persistently wild. Keep a close eye on your tools, adapt them to your needs, or adapt yourself to theirs. Technologies do not subject themselves to what we wish them to do, but interfere with who we are (Mol, 2008: 50).

Further, she establishes a compelling connection between the meaning of work, which one must recognise oneself, and the completion of a product. She believes that one's own conviction is necessary for later potential success.

As with Viktor and Susan, one's conviction generates the purpose—this is also evident in the narratives about the prototypes or products. Conviction or belief itself has several objectives. First, one's conviction is necessarily a prerequisite for being active. It requires overcoming uncertainties to take an active role altogether. It becomes a vital function for surviving the process of innovation, and in addition, it can mean the conscious adoption of responsibility.

Johann, the founder of a company that develops hydrocephalus valves, tells me that the basic prerequisite for developing an idea is a firm belief in its functionality. He mentions that his professor's idea immediately convinced him; at that time, there was no model or prototype but purely physical knowledge about the general functioning of valves. The idea was to make these functions fruitful for the so-called hydrocephalus.

> J: I didn't become a passionate engineer in the first place, but that's how I learned in my studies how wonderful the engineering profession is. Although today I am very fond of it. And that's also how I felt about becoming an entrepreneur. I wasn't afraid of it because I felt relatively secure in my social status. I've already mentioned my economic calculations, and I thought others would go under, but I wasn't so threatened by having 100,000 Deutsche Mark in debt. I think it will be all right, as an engineer you will find a good job. Yes. And then, meeting people, I often succeeded in discussing things that people asked about instead of saying: "Oh, I won't talk to him any further." So, what is the reason now? We have to ask them. I can only say that I think I was convincing. I was convinced, and I could answer questions. I was able to, and I also, I stood up for the cause. [...]
> I: You also talked about self-confidence, and before that, you talked about risk. Does this self-confidence mean belief in the product?
> J: In the idea.
> I: In the idea. Just in the idea?
> J: Yes, I would say now that it might even be less than the idea, actually, in the approach that already came from this professor.

I: You already believed in that?

J: Yes.

I: Would you say that believing in one's idea plays a big role in innovating in general?

J: I think so. Did Steve Jobs know that an iPod, iPad or iPhone, before it was developed and shown on the market for the first time, would be successful? I would say his answer would be: "Yes, I knew it." So, he may have known it, which he has not tried. Unless he knows what others don't know. *(Interview from 11/02/2020, Johann, CEO of Hydro, own translation of the German transcript)*

Thus, the sequence of feelings in this process oscillates between one's own experience and the development of an idea as one's conviction of its necessity ripens.

While developing, uncertainties can arise, e.g. through the anticipation of external doubts or because problems occur. As a result, an idea, a model, or a prototype is adapted, also to avoid further criticism. The conviction and belief in the necessity remain, however, whereupon a narrative is developed for the product. The narrative serves to manifest one's own conviction and, beyond that, the conviction of others (subchapter 4.2.2). Situationally, the narrative can be adapted to be convincing and strong. At an advanced stage, I can generate a form of 'knowledge' by transcribing my belief into a narrative, and this is when I can take features of my previous work as given without assuming that I cannot fulfil them.

As previously discussed, religious parallels are not far off the beaten track, especially when talking about meaningfulness, faith, and influence.

This is a matter of unselfish surrender since innovation is an uncertain undertaking that often fails. In this respect, a parallel with Georg Simmel's description of religious belief can be seen, if not entirely, at least in part. It is also true for fervent desire, humility, exaltation, sensual concreteness, and spiritual abstraction (Simmel, 1905). These Simmelian descriptions of the tension of religions also occur in the belief in one's own idea that is to be successfully realised, producing an ardent desire. Innovators can be humble – as we will also see – in adapting to external circumstances, including problems of implementation and feasibility, criticism, and financing. At the same time, moments of elevation take place, e.g. through unbreakable faith, manifestation in the narrative of the same, and convincing others. The sensual immediacy is the play with my idea, the constant confrontation of my perceptual world with the outside, and the non-sensual abstraction is the attempt to realise all that idealism in the real world despite all the adversities that come one's way.

In one way or another, both Viktor and Johann are convinced of and dedicated to their idea. Both Viktor and Johann are convinced of and committed to their idea in one way or another, whether it is the moral influence, as described, involving the obligation to contribute meaningfully to society or Johann's dedication articulated in the interview. The latter also holds moral significance, as it can be inferred that complete commitment and discipline would not be directed towards an endeavour that did not align with established rules or norms, particularly if it served no further purpose for a group or society.

Figure 13: Hydrocephalus Valves in Different Stages of Development

VI. Premises and Other Problems

Once an idea has been conceived, the innovator or the team decides on a first draft. This marks the commencement of the practical innovation process, which often befalls hurdles and problems. First of all, particular prerequisites need to be clarified, such as the environment in which a prototype is to be developed, namely a makerspace or incubator. Financing also matters – is it feasible to fund the innovation with personal resources, or is external financing necessary? Who is involved in the innovation process, and at what stage?

In the following, I examine the most common prerequisites concerning problem situations during the innovation process. Again, insights from the various interviewees will serve as a basis for the analysis. As previously observed, interpersonal issues tend to be consistent across different environments, primarily due to their emotional impact.

6.1 Finding a Lingua Franca

With the appearance of several actors gathered around an idea and a resulting prototype, communication difficulties may arise. As delineated in detail in subchapters 3.3. and 3.4. several factors can underlie this problem. Hence, there is the hurdle of communication and mutual understanding, either within a team or within the incubator or vis-à-vis the financier. Such barriers may stem from various factors, including professional or ethnic backgrounds, as indicated by my interviewees. Removing these impediments to collaboration is imperative.

The group's gathering, which is defined through its cooperation during the prototyping development process, matters on several levels. First, they need to identify common goals. As stated, the things society explores and the parameters under which it looks at them are contemporary, and the conclusions of a knowledge process are a temporal testimony. This dynamic extends to the language adopted and agreed upon by the group. Finally, an economy, as a collective entity, negotiates the common frame of reference, determining what is permissible and impermissible and whether a shared logic, congruent with moral principles, can be established or, if disparate, external logic must be accommodated.

Ultimately, potential users, not necessarily purchasers, are also an important part of the innovation process and must be included in the communication process. For some, potential users provide a data basis for development. A successful communication process enables people to engage with each other despite having different technical or professional backgrounds. The interviewees report how difficult it is to find a common language.

In a *Believer School* workshop titled 'Réflexions par les machines'[1] [Engl. Reflections through machines], I experienced an extreme situation that exemplifies miscommunication. In this workshop, groups were asked to present their common ideas, visions, or prototypes. The groups came together through an upstream process by formulating their interests on slips of paper, which, for instance, named a theme or also the technical implementation. Subsequently, common interests were identified, allowing the formation of groups comprising two or more individuals.

As *Figure 14* shows, the two participants jointly presented an idea with their shared focus being 'protection', leading them to collaborate on developing an app. The team comprised one female tech artist and one male journalist, as per their self-introductions. The idea was that their app would automatically record a conversation or ambient sounds as soon as you mention a pre-set password, subsequently uploading the audio recording to a designated platform.

When the two were asked about the rationale of their idea, a somewhat perplexing scenario ensued. She confidently introduced the idea as a prevention tool against harassment, coercion, or even sexual violence. When she expressed this, he was totally perplexed and confused, his face contorted before breaking into laughter. It dawned on him that he had an entirely different concept in mind, prompting him to exclaim, 'This is not my idea'. The whole group had to laugh. Unaware of his confusion, she flinched. He aimed to contribute to topics related to 'hacktivism', as he described it, envisioning the idea as a mechanism for uploading specific information directly onto another platform, providing journalists with unfiltered material for the community. Evidently, at that moment, he did not align with her presentation. However, what others found amusing ultimately serves as an example of poor communication. It was evident that the two had failed to clarify the purpose of their idea beforehand, resulting in them developing their 'common' idea in entirely different directions.

Later, I inquired the tech artist about the origin of this misunderstanding. I found it difficult to comprehend that they had not previously discussed the purpose of their idea or engaged in initial collaborative brainstorming sessions. She then said

---

1   This workshop was initially supposed to be in French but took place in a mix of both languages: French and English.

that their common keyword had been 'protection' and that they both agreed on developing an app. From there, they continued to think of 'protection from authority' and 'protection from outside exertion of influence'. After that, they immediately started working on the idea, each having a concrete purpose in mind but not concretising it. They later agreed that 'protection from outside influence' became the common vision, but it was neither referred to nor did they develop a common frame of reference. In her eyes, it was clear that the outside influence was (wo)men ready to use violence; for him, the reference was his own professional milieu and the problems journalists face when they leak information. She later added that they initially kept their brainstorming process as general as possible so as not to limit each other. However, they eventually agreed that they should have been more articulate and clearer about what they wanted to work towards.

Figure 14: Explaining the Idea – Finding a Lingua Franca

While this example of failed communication appears extreme, Johann also describes similar situations concerning his everyday work. The CEO of *Hydro* underscores the significance of being able to express oneself clearly and elucidate product features comprehensively to ensure the recipient understands them. He stresses the importance of avoiding arrogance and instead focusing on aiding the recipient's comprehension. In fact, our conversation served as an educational session for me as he elucidated his product. Ensuring the company's product functionality is comprehensible is paramount to him. He delineates communication's pivotal role as a prerequisite for achieving success.

> So, it's not about being a know-all, [...] but also about always explaining what you mean and then convincing others. Yes, communication, for example, also plays a major role. Not talking past each other. *(Interview from 11/02/2020, Johann, CEO of Hydro, own translation of the German transcript)*

As discussed in subchapter 3.4. and highlighted by Ludwik Fleck's *Denkkollektive* and Lorraine Daston's *moral economy*, (disciplinary) origins are subject to specific logics of values and norms, which are equally reflected in their understanding and their language. When Johann talks about the fact that, as a source of inspiration or product developer, one must make an effort to express oneself clearly and understandably, this is what he means. Assuming you will be understood regardless is fundamental to mutual understanding. His frequently observed problem, therefore, relates to 'talking past each other'. It is imperative for him to avoid this. The process thus includes the effort to meet on a communication level and to gain a shared understanding of the mutual expectations, norms, and values, as well as to comprehend the emotional world and thus the other's judgement logic.

Felix, the external consultant at the clinic incubator Health Hub, also describes situations similar to Johann's everyday work, which mirrors his experience. He tells me about the difficulty of finding a precise language that others understand.

> Well, it was brought to my attention a few times now that maybe I ask too many questions or form a virtual circle of chairs. Last year, I tried to be clearer and said, "Let's do this by next week!" That went down quite well with some people, surprisingly enough. I didn't enjoy it that much because I don't see myself in that role. And we swing into it, I think. It's a mixture of making suggestions, "Hey, now this and that would be good", versus sharp announcements that I don't make and explaining necessity. A lot of it really builds on each other. You can't build or finish designing or programming an application if you simply haven't talked to potential users yet. That simply doesn't work. *(Interview from 13/07/2020, Felix, Consultant at Health Hub, own translation of the German transcript)*

Unlike Johann, Felix grapples with this process, finding it challenging to align with the language he employs. He acknowledges the potential effectiveness of a 'commanding tone' but confesses that it does not resonate with him. The role of issuing clear instructions does not feel natural to him. He yearns for a more organic unfolding of his understanding rather than a descent into forcefulness. It is a revelation for him that a distinct task outline seems to be a more effective tool for project advancement.

Interestingly, some of the physicians I spoke with felt the opposite way. They describe Felix's manner as too dominant, which is said to have already led to occasional conflicts. Bahar, in particular, describes how much pressure she felt from the language used and thus turned away from Felix as a consultant for her project. More on this will follow in the subsequent subchapter.

However, it becomes clear that no common language was developed between Felix and Bahar's team that could have yielded fruitful results. The differences led to conflicts that ultimately ended in Felix no longer being brought in as a consultant for the team. Bahar perceived a lack of serious consideration for her role as a female team lead.

Hendrik, Bahar's husband, is the executive officer of the same project and reports on the hurdles regarding the presentation of results and reporting requirements because the project is funded by public money. Significantly, finding a mutual language plays an indirect role in this.

> **H:** [...] There is frustration from time to time, and then there are somehow evaluation meetings that you have to have because they are for public money.
> **I:** What do they look like?
> **H:** You sit down with the incubator management. That's three people in principle. [...] And then you have to report to them what you have done. You have to prepare the PowerPoint together with the team, yes. You have to show the status; then you go there, then somehow you get the milestone plan back, yes. But you have long since deviated because you have to somehow make progress and you have to report on it, then there are questions and back and forth and because, of course, it's public money, and they have to make sure (for the taxpayer) that the money is used properly. *(Interview from 03/02/2020, Hendrik, Executive Officer for Feety at Health Hub, own translation of the German transcript)*

In this excerpt, although Hendrik does not discuss finding a joint language as such, he describes the process of presenting results to the incubator management and making them understand the status and objectives. As he mentions later, the team tries to communicate so that objectives are mutually clear. Initially, it is still about setting preliminary goals and developing a plan that describes the first processes. In the process of presenting the results, as Hendrik describes it, it is above all about performance and the adequate presentation of the work process so far so that the work is accepted and, at best, allowed to continue in the way suggested by the team.

However, in the interviews with the teams from *Health Hub*, it is repeatedly mentioned (subchapter 6.2.). The meetings and meeting milestones with the management can be problematic precisely because conflicts of interest arise. As Hendrik mentions, 'there is frustration'. Departures from the original development plan for the prototype have been ongoing. It is all about feasibility, which triggers frustration in him. Additionally, processes are severely restricted because the money they have spent is from public funds, and the incubator is accountable. Thus, the incubator transfers pressure onto the teams. The incubator is ultimately interested in accelerating processes; feasibility is in the foreground, often resulting from using unrelated languages and interests that follow different reasonings. The situation Hendrik draws attention to is ultimately one in which the team is careful to speak each other's language. Hendrik later notices in our conversation that, over time, they develop a joint vocabulary. He seems amused that he now uses words he has not heard before working for the incubator. As Johann correctly notes at the beginning, the team must be able to express itself understandably and clearly to succeed. Thus, in this situation, the team is not only accountable for the developments related to the spending of public money, but for them to emerge from the situation with as little conflict as possible, they need to speak the manager's language.

It is different for Ryan, who enthusiastically talks in the interview about how effectively he works with the company that builds his prototype and mentions how well they understand him without him having to explain too much. He is thrilled that they understand each other right away, can implement what he has in mind with his idea, and can communicate without many words, even though they come from different disciplines.

> Well, what's, I guess, the major advantage with [name of tech development company] working with these people is that I just have to tell them one thing once, my idea, and they can actually automatically turn it into reality. They, I don't have to explain to them in-depth what I want, and they intuitively know what to do. This is compared to, maybe, other groups where you have to keep explaining things. "No, I said I wanted this. I said I needed this specific way." I just need to tell them we're kind of on the same page. They just come at it from a tech angle. I come from a medical angle. [...] What's going to be the biggest challenge most likely is that when we have to start doing in-depth patient tests and healthy volunteer tests, working with Shahaf [developer], who's more of a scientist and kind of explained to her, okay, here's the end result of the test we want to have. I think she also comes at it from a different angle. I think she's been in med tech start-ups. And so, she's really just like, okay, we should just go full-on into the software. We should just do this and this and this right away. She's also Israeli. So, she has a very, let's say, different cultural way of dealing with things. Very, just like "we need to do that." So, "we need to do this right now." I'm more from a Scandinavian background. So, I like to think things through a bit. So, that may be a challenge. When saying, okay, we need to do patient tests, we're going to get the data, we'll have to analyse the data. And I think she'll have a different, I think she'll have a different way of analysing the data. We'll have to see how that works out. When I tell, like, the [incubator],

like, okay, in this project, we need to have, here's the idea for the project we need to, we kind of already saw it through how we want to have the end product needed to be financed. I think the [incubator] also they'll say like, okay, well, could you use this for neurology somehow? Could this be used for strokes? And you think probably not. I mean, there's a lot of stuff there for strokes. *(Interview from 04/12/2021, Ryan, Physician & Innovator at Health Hub)*

At another point in the interview, Ryan speaks of an intuitive understanding that makes working with this company very effective. The common language did not have to be developed in this case, ultimately due to the experience of the company's developers working with physicians and non-developers and, further, Ryan's affinity for the technical part. As described earlier, he feels tremendous enthusiasm for technical feasibilities and prefers to work in research rather than with patients. As he mentioned, technical understanding is instrumental in his work, but also beyond that, he says, in a technologised world. However, there is more to Ryan's interview excerpt regarding his perspective on language and understanding. When he talks about his developer, Shahaf, he sees potential challenges in approaching a problem.

As he describes it, he can imagine Shahaf reacting very quickly to possible problems and trying to eliminate them with a software solution. Ryan himself, on the other hand, talks about thinking challenges through before aiming for a concrete solution. He attributes this not only to his disciplinary background but also to his ethnic background. Shahaf, Ryan says, is Israeli and, therefore, he thinks, is more straightforward than he is. He has more of a Scandinavian background, which makes him, so he explains, more cautious in his approach. In Ryan's case, his initial experiences working with his recent developer are mixed with assumptions.

Even though he compliments his developer, he can imagine their approaches are very different due to various factors. In the further process, however, he informs me later in an informal conversation that they have come closer through their collaboration and continue to learn from each other. They do not always agree, but they begin to develop a common path and, further, a common vocabulary. It will even go so far that they develop neologisms that they refer to in their collaboration and with which they begin to identify their work. At the end of this excerpt, Ryan also mentions engaging in conversations and experiencing misunderstandings with the incubator management overseeing the development of his project. Above all, he sees a problem in that different expectations arise regarding the idea and a potential end product. What bothers him is that the incubator is more interested in getting as much use out of the idea as possible, regardless of whether it is feasible. From Ryan's perspective, his idea has a specific scope of application that he believes cannot be easily expanded. He deems the additional possibilities proposed by the incubator to be unfeasible. As a result, the discussion situation is, at times, deadlocked. The incubator thinks about economic utility potentials to gain as much security and, consequently, financial capital from the later end product. Ryan thinks primarily about implementing the initial idea independently of capitalistic profit potentials.

In all cases, finding a lingua franca is a crucial element in joint technology development. One could break these aspects down into teamwork rules, which would be too simplistic. Successful work, meaning work that is not at risk of failure, should insist on becoming a moral economy that shares a language and makes the group's value understandings transparent. To some extent, as we have seen with Hendrik, one part might adopt the vocabulary of a discipline faster than the other way around. This observation points to hierarchies that can evolve. Especially concerning different ideals or ideas of success, the projects seem to aim for feasibility, especially towards the end of their duration. Focusing on feasibility seems pragmatic and yet does not seem to justify the idea's origin.

6.2 Conflicts and Emotional Decision-Making

As demonstrated in the preceding subchapter, conflicts are not mere hiccups but significant hurdles that often stem from different languages and communication patterns. The presence of a joint language can help avoid these conflicts; however, in its absence, a myriad of emotions surface in these conflictual situations, influencing crucial decisions in the development process. These conflicts and disputes are pivotal crossroads in the process of innovation development, carrying a heavy emotional weight.

The innovation process is a complex interplay of benefits and challenges, all stemming from its inherent diversity. In this section, we witness the diverse actors and their unique perspectives attempting to overcome obstacles and unite as a cohesive whole. While a common language as a tool for effective communication may seem like a straightforward solution, it underscores the intricate nature of this unification process. The data material also reveals that problems are not predictable but rather emerge from the richness of this process. And this multiplicity, it transpires, has a profound impact on the team, the cooperation, and the development of the idea. The material, therefore, offers a glimpse into the daily dynamics of innovation in any setting.

In the following, various excerpts underline the emotionality in conflict-ridden situations. Bahar starts by saying that she spends 90% of her time solving problems as a team lead. For her, a doctor, these are not only unfamiliar and new tasks that otherwise have nothing to do with her work, but she also mentions that she often feels disoriented. Due to the novelty of the tasks and the unrelatedness to her previous experience, she frequently encounters problems that necessitate individual solutions.

Well, we are constantly making experiences, whether in the hospital or here at Health Hub. And then, not only with the product but also always in the team, with the people; probably everything flows into the work somehow. I just know that good experiences are pleasant, but mostly, the bad ones help us innovate. Because then we know what we must change and what we still need to do. In brief, bad experiences are the ones that make us think, and they are easier to sell as a result. [...] My job is to keep the team together, solve problems, keep distributors in check, [...], I mean, 90% of the time, you have problems like – sometimes my "male problems" sit next to me in meetings, and I don't let them talk, so everybody in the room knows I'm the boss. So, those kinds of problems. [...]

But that's – oh, I've got two team members who are both over 1.9 m [tall], and whenever we go into meetings [...], the others always think one of them is the shark, and then the little dark-haired [she speaks of herself] starts to swear and that's just always not a model for everyone else to get on well with. Most of our service partners are in their mid-50s and have been working with the same partners for 30 years. And that's just how it is sometimes, unfortunately. These are the problems, and it's always so, how shall I put it? It's always unknown territory like my patient has a pulmonary embolism, and I know what I have to do. It's not like that, but somehow, everything needs an individual solution. *(Interview from 30/01/2020, Bahar, Physician & Innovator at Health Hub, own translation of the German transcript)*

However, Bahar is even more disturbed by the fact that conflict situations often arise due to misunderstood hierarchy. She feels discriminated against due to her gender and not taken seriously, which she often describes as a conflict situation in the interview. In this excerpt, she refers to the fact that she is not taken seriously by male service partners who are approximately 20–25 years older than her. She describes both Viktor, the team's developer, and her husband, Hendrik, as two tall men who, in her experience, are more often perceived as team leaders because of their phenotypical appearance and gender. In this context, and due to the feeling of being pushed aside, she relates that she has exhibited a certain behaviour whereby she becomes dominant in appearance and speaks loudly and brashly. It is the moment when problems in communication with others and the development of the product become apparent. These conflicts are influential in that Bahar often attributes them to gender differences. As a result, she has changed her appearance and behaviour. Further, she decided to refuse to work with Felix, the incubator's commissioned consultant and his team.

B: Another problem was that they always tried to put the young colleagues [from the consultancies] into the teams, so to speak so that they would do it [consult], and sometimes they didn't even understand what it [the project] was all about. And that was, well, that was the combination of these things that made it difficult.

I: Okay, and then Felix came along, and things got better?

B: And then he came along. It didn't get better for me. Because he didn't recognise me as an authority figure in any way. So, I didn't get along with him. It went so far that I called [Leif (chief physician and Bahar's supervisor)] in because I thought he wouldn't listen to me. And then [Leif] came along, and then he put him in his place, and then everything worked. Everything has advantages and disadvantages. [...]
I: It's based on sympathy?
B: Very much. On the other hand, I have to say, we also had advisors here; I don't know if you know [Basil]. With [Basil], every five minutes of conversation has been efficient. That has always brought us further. Just like with other consultancies here like Johner [medical advisory institute] or something. That has always been effective. [...] But as you say, with other things, it was sympathy-based, and then it was also something different for us initially because we had a dependency on the technical developers in the first round. The consultants hired the ones who built these sensor things. (I: I see) and so we were in a very isolated position here because, for everybody else, they just did a bit of consultancy. For us, they actually did product development, and we were totally dependent on people who didn't like us. *(Interview from 30/01/2020, Bahar, Physician & Innovator at Health Hub, own translation of the German transcript)*

She explains her decision to quit working with the consultancy as being due to the feeling that Felix was too dominant and that he did not listen to what she said or consider her experiences to be valuable. She felt so uncomfortable working with him that she needed to ask her supervisor to join the meetings. In this context, it should be mentioned that the supervisors usually do not work in the incubator or with the team but rather fulfil their obligations in the hospital. They are listed as supervisors on the application form at the beginning but leave the teams to themselves. Leif's appearance changed the dynamic, but this was not a solution, as the atmosphere was not sustainably improved.

Nevertheless, during our conversation, I notice that she also insists on the hierarchies she criticises at the same time. Among other factors, she perceives a disadvantage due to the assignment of young colleagues from the counselling team to her. She thinks she has to explain her project more and that the young consultants have little idea about her project. At this point, she seems annoyed. What further exacerbated her discomfort was the consulting firm's establishment of contacts with the technical developers initially assigned to develop a prototype before Viktor joined the team. The consultant's placement at the development company gave Bahar the impression that they disliked her and her team. This remains unverifiable. However, her impression and feelings of being held back and lacking recognition led her to refuse to work with the consultancy firm. The conflict also does not leave Felix unscathed. Whenever he talks about communication problems, Bahar's team comes up. Later, I learned that Bahar had left her team during the COVID-19 pandemic, and so did her husband. Although he does not meet the incubator's eligibility criteria, Viktor remains alone as an external employee who is de facto the team in the personnel union that drives the prototype further.

The conflicts Bahar describes above are also noticeable within the team. Viktor occasionally indicated in our interviews that conversation situations were often conflictual and that he frequently saw the project work slowed down due to mismanagement.

> **V:** [...] When I'm going back home to Romania, you know, before I go, I'm trying to give everybody work, like to have things [results] afterwards, and I was always trying to do this. And things didn't always work. There was a huge delay from the measurement company—there were always delays that were not my responsibility or that I could do anything about.
> **I:** Do you find this frustrating besides being fascinating?
> **V:** Frustrating, sure. Like the fact that I didn't get it done before I left Germany for vacation. It was frustrating, yes, of course. And, looking back right now, it was obvious that we should have bought, like, a piece of equipment that cost like 4,000€, and we didn't because the team lead didn't see the necessity. And this can like... [it] delayed everything because we were dependent on this company for measuring the load cells. Like – looking back, there were some big mistakes that delayed the whole thing for months. Because when I arrived here, I said we need that equipment. But no one listened.
> **I:** Is this one of the reasons you have sometimes this tense atmosphere?
> **V:** By tense, do you mean the fights we have? *(Interview from 04/02/2021, Viktor, Developer at Health Hub)*

Viktor often feels discouraged as a product developer who brought much experience into the project through his studies and previous work experience. He describes how mismanagement and unproductive discussions in team situations lead to unfavourable decisions being made for the project. In the conversation, he tries to exemplify this with a situation. First, he describes that he usually tries to distribute tasks before he goes on vacation so that he can continue working after his return, at the point where the other contributors leave off. This time, however, he did not manage to do so, partly because he depended on the company that manufactured the weight sensors. Thus, there was a cascade of delays that annoyed him.

On top of that, he resented the lack of equipment that he thought was necessary, but the team lead, Bahar, did not. In retrospect, he sees significant mistakes here that disrupted the development process and led to team arguments. In fact, this quarrel was a situation I witnessed during a visit, which is why I asked so specifically about the tense atmosphere. On the day in question, I had an appointment to interview Bahar. When I arrived, I was told to wait in the corridor while Bahar, Viktor, and Hendrik finished a conversation. They were arguing, and a door banged at the end, with a murmur from Bahar.

Viktor occasionally attributes these decisions to Bahar's inexperience as his team lead. Later, he also talks about different types of communication and considers Bahar's manner aggressive, although he emphasises not holding this against her. Conceivably, his observation only confirms what Bahar said about herself as changed; dominant behaviour should indicate that she is the actual team lead.

> [...] I think that happens often with us. Like, that's the way you're talking. Some people have a personality, and they communicate in a certain way which is a little bit more aggressive than others. (**I:** Mhm.) But that's not my style, but I can understand where they're [Bahar and Hendrik] coming from. [...] You know, sure, there are consequences. We've already felt them. I mean, we've already broken off work with others, that's not good, but...I'm not the boss; I do things differently, but I accept that. *(Interview from 04/02/2021, Viktor, Developer at Health Hub)*

Viktor elaborates on the extent to which past decisions were emotionally driven and contentious. He rationalises this by speaking of perspectives when he says: 'I can understand where they're coming from' and refers to Bahar's and Hendrik's temperaments and situations. He does not necessarily find the resulting decisions sensible but comes to terms with them. However, he emphasises that emotionality leads to specific decisions. He speaks of the consequences they had to bear as a team and indirectly addresses the fact that the cooperation with the consulting firm, especially Felix, had ended.

Johann sheds light on another aspect that has so far remained unexplored. At this point, he refers to the patients, i.e. the users, who are ultimately confronted with the technological development in everyday life and are on their own.

> After all, our patients are also crucial [for technology development]. They tell us if something doesn't work, so it's all about functionality. So, these are things that you can only check in everyday life. [...] And yes, I mean, how they feel with the valve. If it's not a practical solution for them, or if they don't feel comfortable with it, that's important. And emotionally, because you asked before, so yes, that's emotional.
> They do rely on something, on technology. And you know, suddenly the living conditions change again, so they become better, less life-threatening, of course, it's emotional, what else. *(Interview from 01/08/2020, Johann, CEO of Hydro, own translation of the German transcript)*

Users provide an essential impetus for technological development as certain features may be unsuitable for everyday use. Although this lack of suitability is sometimes only seen in the experience of individual patients, these insights are indispensable for him and his team. The factors mentioned influence the further development of the valves, and, as he says, these factors are (often enough) emotional as they are closely linked to the users' everyday lives. They often give users new hope when, as Johann says, they suddenly create new primary conditions for them that make a new quality of life possible. What Johann describes here, however, does not apply to the same extent to the incubator, which develops its applications for patients. Here, patient data only matter peripherally. Initially, they are evaluated when it comes to identifying and defining the exact deficit of the old (i.e. existing) application. In the further course of development, the team itself tries to determine whether it is suitable for everyday use. Johann describes it differently here, as the patient data seems to be indispensable for development. I could not track the extent to which they were incorporated, and, in general, it was difficult to determine to what extent user data actively contributes to the development.

> In the end, there are many reasons to decide one way or the other. Emotions also play a role again and again, perhaps even always, because if—well, let's say you have, there are financial reasons for a decision, then it can still be emotional, right? Well, emotions are always also between people, but they often enough relate to other things, you know what I mean. *(Interview from 26/04/2020, Christian, Founder of M.lab, own translation of the German transcript)*

Christian gives a general assessment of emotions in relation to innovation in our conversation. According to his assessment, emotions are always at play, whether alone, in collaboration, or concerning other problems. Even if they do not occur in the foreground or are considered a factor in decision-making, they are still part of it. He notes that it does not matter whether an emotion is identified as an impact factor; he is sure they are always part of the process behind the scenes. The innovation process, with all its decisions, is often sufficiently fragile and sensitive, which is why those involved are emotional. Situations *I* assess, people *I* interact with, and decisions *I* take based on my assessment, as situational exchanges of human experience with *me*, others, or about something, are emotional.

For Bahar's team, it is evident that emotions have influenced decisions regarding the prototype on multiple occasions, beginning with her initial anger, which catalysed the development of a solution. Also, resentment over perceived disdain makes her no longer want to continue working with Felix. Alternatively, as Viktor describes, Bahar often makes emotional 'gut-feeling decisions' for the project that he would have liked to have weighed more carefully. However, he is not exempt from his emotions, as evidenced by his feeling of not being heard and his advocacy for a different approach to the project. His feeling of being disregarded gives rise to disputes. In the end, Viktor also makes decisions for the project out of a need for security since he has been working alone. The main reason is that he wants to secure his work financially. In this context, Johann discusses an important point that does not seem relevant to the other teams I observed: what is decisive for him are the patients' emotions as they must live with the medical solution. Therefore, in Johann's eyes, they should have the final say.

6.3 Trustful Coalitions

To overcome conflicts and create the most stable environment possible for innovating and collaborating, trust related to the cooperation with external parties, within the teams, or to oneself was highlighted in all the conversations. As previously discussed in subchapter 3.4, once a community has formed a moral economy, it benefits from its variety and creates new possibilities. However, as noted earlier by Emil Durkheim, for this to occur, individuals must transcend their previously experienced 'mechanical solidarity'. In my research, it is possible to observe how individual team members transition from their original 'archaic group', to which they feel they belong due to their similarity in work, education, and lifestyle, into collaboration in an incubator or a new team, developing a functional 'organic solidarity'. This organic solidarity refers to cohesiveness based on the interdependence of people in increasingly complex relationships whereby interdependence resulting from the specialisation of labour and the complementarities between individuals is a characteristic of 'modern' and 'industrial' societies (Adam et al., 2000; Beck, 1986; Durkheim, 2013).

However, the melting pot we encounter still needs to create organic solidarity and trust from which the group could benefit. In fact, these are fragile processes subject to the sense of belonging and thus determine the extent to which individuals feel a sense of belonging and commitment to their group. I begin by sharing a quote from Bahar that we have encountered before, describing the difficulty of leaving a familiar, exclusive group. The process of leaving the previously familiar surroundings and abruptly entering a new working group in the incubator involves facing a transitional process in which neither one nor the other form of solidarity is felt.

> If you think about it, during your studies, how many medical students have you met? Not so many. Most of them have their own campus, usually, they are located in the university hospital. That's usually on the other side of town. And, of course, you also have a circle of friends that is so exclusive. When I came out of my studies, I didn't know any software developers or technical designers. I didn't have these people in my circle. I just had other doctors. But you can't found a start-up from five doctors, not for MedTech. And then we started here with external contractors. But that wasn't so ideal. They only want the money, and what they deliver is always the minimum version. And then, by chance, we got Viktor, our technical developer. He studied computer science at MIT and has five years of experience in designing wearables. And then we got someone for the business administration part. My sister

is already back in her studies; she dropped out again. *(Interview from 30/01/2020, Bahar, Physician & Innovator at Health Hub, own translation of the German transcript)*

What Bahar describes in this excerpt is the challenge of departing from an 'archaic group', as described by Durkheim, and building a team with members of diverse backgrounds. She highlights the absence of interaction with other groups, the earlier seclusion, and the abrupt shift and adaption to situations that did not exist throughout her medical studies. She claims she had never met a software engineer or technical designer before attending the accelerator programme. The technical difficulty of realising a notion overwhelmed her, and she lacked the necessary abilities. She was required to locate people with the skills she needed and who were willing to begin the process of reciprocal translation with her. Reciprocal translation entails accurately synchronising the scopes of expectation for a prototype. This procedure is complex since it does not begin with the same terminology. Due to the diversity of backgrounds, each instance strongly emphasises what is deemed essential. As shown later in this section, the joint coordinating procedure is perpetually demanding, and she considers locating personnel with exact expectations for a specific task challenging. The external contractors were of little assistance since they had different ideas or, as she claims, 'just wanted the money' without presenting a version to which she consented. Therefore, Viktor was a capable individual who became connected with her and her initiative. He was the prospective team member to whom she wanted to commit her concept since they shared a similar perspective and vision for its direction.

Trust, as my informants emphasise, is not just a tool to build a team but a glue that holds together a diverse group with a shared sense of collectivity. It is this trust among themselves that paves the way for successful cooperation. Anthropological studies (Adam et al., 2000; Corsín Jiménez, 2011; Frederiksen, 2016; Ingold, 2000) have consistently underscored the necessity of trust in maintaining the stability and robustness of social relationships. They also reveal how trust permeates (corporate) knowledge, its underlying culture, and systems of responsibility (Corsín Jiménez, 2011). The team's shared understanding, often referred to as trust, is crucial. Equally noteworthy is how the collective narrative not only binds the team but also elucidates the societal need for the product. These identity-forming narratives, a consensus of values, are essential for collaboration and instil a common conviction in the product, one's own effort, and future reward.

In Bahar's case, corporate knowledge or an underlying culture has not yet been cultivated because of the lack of stability in the transitional phase.

At this point, it is worth taking a closer look at the different forms of trust I encountered in the field. Consequently, I must retrace my steps and revisit some earlier theories. As explained in subchapters 4.1.2 and 4.2.2, trust is both an emotional and a cognitive category. According to their respective meanings, emotive trust may be ascribed to mechanical solidarity, while cognitive trust can be ascribed to organic solidarity.

Assume that I am a member of an ancient community in which I am dependent yet exist based on reciprocity. In this situation, I profit from a trusting relationship built on benevolence and voluntarism. This conception of trust is emotional due to its unconditional nature. Suppose, however, I trust because of artificially established group constellations, the required openness of specific procedures, or any other duty. In this situation, trust is a cognitive category of order. Consequently, trust exists on the surface, even though the ostensibly trustworthy relationship has been compromised.

I will present different interpretations of trust I encountered in my fieldwork. Distinct types of emotional and cognitive trust arise due to dependency alone.

Felix, the incubator's external consultant, describes the importance of trust and the emotionality that go hand in hand. He points to trust as an affective category and tells me how important trust is as a basis for good cooperation and how much joy he feels when this level is reached. He describes it as a feeling of 'togetherness'—a unity that happens over time and consequently due to closeness, belief in each other, and joint work.

> [...] Relationship of trust. That's great and very positive. And I also find it very emotional. Positively emotional. So, it's simply fun. And that is profitable for both sides. And then it also starts to become a togetherness. It's also a relationship that you enter over time. I, at least, enter a relationship for a time. But it can also be negative! That's always when – yes, I would put it down to trust. If the people we are looking after – yes, I called it resistant to advice earlier. This is often coupled with arrogance. With an inability to put one's own personality aside. That can tip over into arrogance. *(Interview from 13/07/2020, Felix, Consultant at Health Hub, own translation of the German transcript)*

On the other hand, he describes how the experience of working together can also be damaging if trust is lacking. He sees the insufficient recognition of his work—a resistance to counselling due to arrogance—as the result of a lack of trust. For him, this kind of situation arises when team members do not take his work seriously or, as he says, '[they are unable] to put their own personality aside'.

What Felix describes as desirable in a working relationship for himself aligns with Ingold's description that was outlined in the theoretical part: 'To trust someone is to act with that person in mind, in the hope and expectation that she will do likewise – responding in ways favourable to you – so long as you do nothing to curb her autonomy to act otherwise (Ingold, 2000: 69–70).' The working relationship should not only be functional in aspects of the work processes but also be human and thus stabilised.

The accelerator programme's leader, Jan, makes a similar argument, albeit from another perspective. Incubator leadership has shifted away from emphasising set criteria such as the competence of external personnel or the acceptance of fully formed concepts. In our second meeting, Jan tells me that human considerations have taken on more significance since the programme's inception in 2018. However, this mainly pertains to interpersonal abilities, which, believed, would lead to cooperative trust. It is more about 'humanising' the programme or mechanised terminology that refers to human skills.

> **I:** What's becoming important instead?
> **J:** To make the teams understand that it is a development programme with many new unknowns; that it is a team sport where a lot of things are already there, so, from the medical-scientific area, but where there is still a lack of technology and, above all, business. And you need all that. Ultimately, I would say that team building, i.e. external companies working with the teams, is becoming more and more important, and well, human aspects, making sure – complementary skills are not enough and compatible time slots-, but in the end, people also have to, I'll say, share values and trust each other to some extent. Otherwise, it falls apart because it is an extra activity for everyone. *(Interview from 13/08/2020, Jan, Head of the Accelerator Programme at Health Hub, own translation of the German transcript)*

In the meantime, Jan focuses on the many unknowns that such an innovation process entails. He indicates that medical expertise is available but knows that the business and technology side does not exist from the start and has to be bought into the team. This is where the team processes begin, and with them, the 'human' impact factors and other unknowns. Thus, the job of 'teambuilding' is added to the objective tasks of 'job fulfilment', and he outlines this as another activity. Expertise and job-relevant skills are, therefore, something that can be checked; however, he attributes the human aspects to trust and to the fact that expected values are shared. Mutual sympathy, cooperation, and goodwill in the team remain untransparent hurdles that must be overcome by those involved. Accordingly, there are predictable and unpredictable factors in developing an idea and assembling a team. Although teambuilding measures can support the latter, the supposedly 'human factor' remains the biggest unknown. However, the question that arises in connection with the unfolded theory at this point is: if trust is the vital and human impact factor that heads, consultants, and teams desire and become indispensable as a skill, can we then start from a concept of trust that is supposed to convey the emotional – sympathy, shared values, a sense of belonging – or is it much more cognitive because it is presupposed and thus cannot be socialised? It is not so much the reinterpretation of a concept of trust that causes perplexity but rather the confusion it creates in the mutual understanding among the team members.

In this context, the different interpretations among the members of Team *Feety* are insightful precisely because the quality of personal relationships differs. Hendrik provides a unique perspective and imparts a particular function to the concept of trust. He describes trust from the team-level perspective, which depends on relying on each other's professional expertise. He says that he and his wife – who have a different level of trust because of their private relationship – only work a few days a week in the incubator with Viktor, the developer. Hence, the team frequently faces the issue of having limited time to address problems collectively and deliberate on subsequent actions.

> I am not working full-time [in the incubator]. Neither is Bahar. And Viktor only has one day a week with us, when we can get together to discuss things. And then he has a thousand questions, a thousand things that come up, where he wants feedback, and then there is always a question, Bahar is the team leader, yes. That means she has decision-making power somehow. But I also see the team a bit more as a joint process, discussion, and decision, so also questions of leadership, actually, yes. What can I lead, how much should I lead, yes? But also, delegation, so when she gives Viktor an assignment to do something. And when he does that, then I also have to ask, okay, why do we have to give feedback so often? I trust him technically because we can't evaluate it anyway. *(Interview from 03/02/2020, Hendrik, Executive Officer for Feety at Health Hub, own translation of the German transcript)*

At the same time, Hendrik looks at the different roles within the team in the context of trust. Again, the ambivalence in Bahar's role as team leader comes to light as she has decision-making power, and yet Viktor is the one who has more technical expertise. Hendrik would like to hand these things over to Viktor to save time during development. He feels conflicted about not overriding his wife, Bahar, as she is still the one who ultimately makes the decisions. However, Hendrik is aware of the tension and wants to give Viktor more freedom in the development process. These conflicts become increasingly acute as the prototype development continues. However, this is not owing to a lack of trust in Viktor but to Bahar's conflicting responsibilities, which weigh heavily on her. Bahar told me in a conversation that she is unsure of her role and does not know how to handle certain circumstances despite her more dominant demeanour. She is also aware of the difficulties faced by the team. Still, she blames her lack of ease on sensitive triggers related to her position as a young female team leader who is not taken seriously. Overall, the squad appears well-managed, and Viktor seems to hold a more significant role than Bahar acknowledges. On the one hand, although Viktor's relaxed attitude is a nice balance, she also experiences a feeling of insurmountable inferiority. Ultimately, it is unknown to what degree Bahar's self-confidence is affected by the disagreement, which may damage team trust.

Ryan is facing other problems with his project, *Ellie*. He is dealing with conflicts regarding patent rights, as an external company was involved in the development before Ryan got accepted at the incubator. Here, attitudes seem to be hardening, and Ryan, as the developing physician, sees himself as a mediator in the legal dispute. His focus on trust at this point differs from those described so far.

I: How do you try to solve these conflicts?
R: Yes, it's a very difficult fine line because I, what I try to do is mostly, I try to just find the people that I trust mostly at [Health Hub] and say, here's the problem I'm having, right? [The external TechCompany is] not seeing eye to eye on this or [external TechCompany] is having this idea. I think this would be a good idea, but they're like [no, Health Hub] not. So, I tried to find a person that I trust there at the [incubator] and say, look, here's the problem, how to best mediate this. *(Interview from 04/12/2021, Ryan, Physician & Innovator at Health Hub)*

Ryan turns to people to ask for advice in a delicate situation. The incubator is his employer, and at the same time, the technical development company has joined in the project for a long time. He thus looks explicitly for people he trusts to find a solution. The problem-solving process is strongly related to believing in the one person he hopes will point him in the right direction as the incubator provides minimal assistance and stability. The incubator would act solely in its own best interest due to the potential for litigation over the intellectual property matter. Still, Ryan does not want to alienate the other tech businesses because he relies on them and has an excellent rapport with them. As he prefers to surrender his job to various organisations, he only trusts certain persons rather than the whole network.

When I ask Karwen about the relevance of trust, he also says that trust is indispensable to developing or financing a product.

I: [...] Why is trust so important?
K: Mhm, there are several reasons. [...] Well, look, I mean, it doesn't work without it. You need a team you can work with, you need people who help you and don't steal your idea, for example, yes? And well, then you need money, so either you have it yourself, sure, but if not, then you need it from others, and they won't just give it to you. You have to convince them of you, your idea and so on. If you can't do that, you don't stand a chance. So, trust is the basis, no? No matter how good your idea is, if you don't get the people on board, forget it. [...]
I: How do you convince people to trust you?
K: (Laughs). That's my secret.
I: Seriously. What do you tell them?
K: Mostly what is. Sometimes what can be? Look, if you need money, you promise

things. A bit like when you get married. You promise something for the future and assume that it will work out. [...] *(Interview from 18/06/2021, Karwen, Private Investor & Innovator, own translation of the German transcript)*

Karwen is an experienced innovator and developer. In his mid-30s, he has already built up four different companies in the past and invested in four more; he is currently building up his fifth company and has recently hired three employees for it. He describes trust as the starting point for good teamwork and the possibility of financing an idea, as one depends on external help, e.g. a business angel. Trust is, therefore, the framework for everything interpersonal, based on belief in an idea, in a person, and mutual sympathy. 'Trust is the basis.' He also explains to me that these are often 'advance praises', i.e. trust granted without guaranteeing success. It is about good intentions first, but he also indicates that failure can be an option, like 'when you get married'. Although things are usually promised with good intentions, these promises may nonetheless not be kept in the future.

Various conceptualisations of trust emerge, characterised by inconsistency and lack of familiarity. They often appear solely utilitarian to advance a project. However, what occurs when a relationship purportedly based on trust proves untrustworthy? The project acts as a projection surface for all parties for an extended period, when expectations are high, and reality might surrender to a pretended relationship of trust. The gap between the desired outcome, fostered by (functional) trust, and the actual effect increases with time.

6.4 'Fake It Till You Make It'

Innovation circles often postulate that failures are inherent to the innovation process and may even pave the way for future success (e.g. Farson & Keyes, 2003; Higgins, 1975; Wills, 2019). As discussed in detail in subchapter 4.3.3, innovators commonly assume that failure is part of innovation practice. It is an optimistic interpretation of failure, which, however, means compulsory survival practice. At first, this possibility is present to the innovator in every early idea, but it is initially faded out. As Karwen indicated earlier, the intentions are usually very good. (Self-)trust, conviction, and the sharing of a narrative about the idea are the tools to create a framework that allows the development of an idea.

The framework can, therefore, be an incubator or the financier, i.e. the business angel. As already noted, my interview partners tell me about different motives for their ideas. However, these motives often do not appear later in the narratives. The narratives change throughout development, as will be shown in subchapter 7.4, since the interviewees adapt their ideas and narratives depending on whom they need to address and convince (see theory in subchapter 4.2). Narratives and underlying beliefs originate from envisioning a particular future, which, as previously stated, aims to address a problem needing resolution. In this process, individuals generate images to stimulate the imagination. Due to the different worlds of experience of the individual, target groups react differently to images conveyed through a narrative. In this respect, it is crucial to decode these worlds of experience and to find a suitable image that, in turn, stimulates the imagination. In brief, the adaptation of a narrative depends on several factors. It is the iteration loops of the prototype whilst the artefact is still subject to ongoing changes. This developing corporate culture helps constitute shared values that stabilise the team or the person the innovator wants to address situationally to convince her (subchapter 4.2.3). The narrative of a project or a founding myth are equally stabilising factors that contribute to legitimising a problem solution: the narrative acts as a framework that delineates and consolidates shared values within the team, serving as a symbolically charged medium of translation capable of adapting to its audience to externalise the values established by the author. Obtaining the desired legitimacy remains a vague business, and often enough, pressure and deception are part of this campaign and are used to steer narratives in specific directions. The problem that arises from this is the traceability of feasibility and, thus, as a consequence, the innovation itself (subchapter 4.3.3). The fact that innovations fail more often is no longer a strategy for achieving 'learning success' but the inevitable consequence of an over-optimistic narrative.

The first excerpt from the interviews related to this aspect is about Bahar's initial attempts to develop the sensor sole and, building on this, her application for the incubator. She portrays to me how, based on her previous disappointment and anger due to a lack of insoles for her patients, she tried to make a sole herself and was later referred to the incubator by her supervisor. Neither the sole nor the idea was fully developed when the application was submitted. She admits to having invented the application's content and the promises without guaranteeing success. She speaks of the application as 'a tissue of lies' in the hope of placating the expectations raised.

> And then I thought, okay, apparently there are no other soles. And then I looked at the soles I had and thought, OK, they don't look that expensive. I filled up the individual parts and came up with a purchase price of 40 euros, and the soles cost 2,000 euros over the counter. And then I thought, even as a doctor, you have a bit of IT knowledge. My sister is super enthusiastic about IT. And then I ordered the things from home and thought maybe she could build them. And then my boss came and mentioned something like "Oh, the [incubator] and money and grants." Then he said, "Ms [Bari], apply!" Then I looked at the call for applications and said, well, that's not really research money. It's about start-ups and funding. And the boss again just said, "Money is money. Go ahead." I called here, and they said, "We don't have anything yet. But we might be able to get something built." And then, they said to me, "You have to submit the application by May. You can formulate

what you think you could have by then. And you don't have to present it until June."
In other words, the entire application was a tissue of lies. (laughs) *(Interview from 30/01/2020, Bahar, Physician & Innovator at Health Hub, own translation of the German transcript)*

Another problem in this context is outlined by Ryan, who works at the same incubator. His product is far from meeting the expectations described on the product website, and his problem lies in the different perceptions of the product based on commercial expectations, among other things. He states how often he has to discuss with his incubator that it is a very specific product that specifically addresses the problem of anaesthesia. However, the current product website also claims that Ryan's idea addresses and covers several medical areas and problems. In our conversation, I dig deeper. The discrepancy between expectations, promises, and the development status quo could not be more visible. However, it also becomes clear that Ryan is not the one proclaiming these promises but that it is an instrument of the incubator to generate greater interest on the outside, which in turn should create a larger market. Whether these are empty promises at this point or whether the research is to be steered in a specific direction will be examined in more detail later on.

> **R:** I think sometimes the [Health Hub] may say: "Oh, but could you use it for neurology, somehow, could this be used for strokes?" And you think, nah, probably not. I mean, there's a lot of stuff there for strokes. This is for anesthesiology. […]. Sometimes they […] envision it in a different way than I do. I think that comes from not working as an anaesthesiologist. They just may not have the needs of that in mind.
> **I:** Would you, but on your website, for example, it says that it's also usable for different areas, right?
> **R:** Could be applications. Yes. Could be. We have to test these out. So, there are a lot of claims, so we have to test them out.
> **I:** But is this due to the [incubator's] wish of using technology? Yes. Also, for different areas?
> **R:** This brings me to a very good point […]. That brings me to a good point because the [incubator] wanted, they want to make things look as dramatic and robust and saying, "this is the be-all-end-all for all products, this is going to be the best thing ever." We don't want to end up in a Theranos situation. I don't know if you're familiar with that. So, we don't want to say this is going to be the thing we can measure. It could be that when we start really testing, it could be, it doesn't work at all. It could be, that is very possible. It could just say, hey, it measures some signals, but in the end, they have no meaning. It could be, it very well could be.
> **I:** Is it, I thought you were already testing regarding…?
> **R:** We tested just a handful of patients to see if we get signals. So, the patients wore

these little sensors, and they were giving signals, but we don't know, do these signals actually have meaning, do they actually come from the brain itself? That's what I mean, maybe these pulse waves are just being reverberated from somewhere else in the body. And they have no correlation with brain flow at all. Because the problem is you can't really stick a probe inside and see actually what's happening. We're kind of, we're making a claim. We think, this is measuring the blood flow to the brain non-invasively through these sensors, which are basically taking in signals for the skin, the scalp, the skull, and the brain. We're making this claim. We have not proven, maybe I should have said this before. This is what we aim to do, but we have not proven that this is actually [the case], so it could be that when we do these, these tests on 50 patients and about 50 healthy volunteers, which should be happening in 2022, we can actually get enough data to show, okay, it's doing what it says it's doing. *(Interview from 15/10/2021, Ryan, Physician and Innovator at Health Hub)*

In our conversation, I find myself somewhat perplexed by the unproven nature of these possibilities. On the website[2], so far, they are presented as established successes that have undergone testing and offer prospects of a certain outcome. Nevertheless, the testing is not the case. Ryan explains to me that these possibilities and options lie in the future and could be proven, but de facto are not. He tells me, however, how eager the incubator was to formulate as 'dramatic' a story as possible from the idea that would achieve the greatest possible attention potential. It also becomes clear at this point that developers and the incubator partly overlook the user perspective. Although there are a handful of tests, they are not yet of any particular value, as the data is still too limited to make any statements. However, the patient value that exists is of hardly any importance at this stage of development. What matters so far is to sound out the potential market success. Ryan is also sceptical: he does not want a 'Theranos situation', as he says, and he would prefer to be able to develop and test in concord, explicitly for the area of application he has in mind, quite independently of other fields of application. At this point, however, his idealistic vision is thwarted by various expectations within the incubator.

Ultimately, the observation remains: 'We're making a claim. We have not proven that this is actually [the case].' Ryan lacks data to give validity to the claims. Although they have test data, they cannot say which pulse waves were measured and whether they are significant for the prototype. Ultimately, the incubator assumes responsibility for the functioning and non-functioning of an idea. However, it is imperative to question to what extent a public incubator breaches its responsibility by assuming or publicly proclaiming prospects of success that have not yet been proven. No

---

2   On 2 August 2022, I noted that the website has since made a change and explicitly mentioned that, until now, it is an idea rather than an available solution.

potential damage is to be assumed here, but, in the worst case, the lack of credibility remains for an institution that works with public tax money.

Karwen's description is different: he does not develop in cooperation with an incubator. However, he raises his money privately via business angels or on platforms like IndieGoGo or Kickstarter. He thus enters into a different dependency, in which he also bears much more responsibility for his promising product narratives if he wants to convince a backer. In this excerpt, we are still talking about pledges and the responsibility of an innovator if he wants to raise external funds. I also ask him about the verifiability of a good idea. He comes up with the much-used adage, 'Fake it till you make it'. This saying is a common expression in start-up and innovator circles, which means keeping a straight face in the public eye or the narrative until success is finally achieved.

> K: You have no idea how much people lie (laughs). But something like that, yes, of course, no one says that out loud, but basically, [...] it's an unspoken secret. But people don't do it because they want to steal money from you or something, that's clear, but out of necessity, I'd say. I've already said that. Well, you know there's this saying you always hear: "Fake it till you make it." I mean, das kommt nicht von irgendwoher [sic! Analogously: "this is not a coincidence"], as you Germans always say (laughs). [...]
> I: Yes, well, but you can also exaggerate, you know. So, am I, as an investor, supposed to know how long I can rightly believe in something, so something is justified hope and at what point it might ultimately just be desperate attempts to keep a bad idea alive?
> K: You can't. (I: Not at all?) Well, you can try, but you need time for something like that. You would have to invest a lot of time to check whether an idea is worthwhile. (I: Mhm.) And in the end, this time might be wasted because you can also pay, right, in the hope that you'll get more money out of it. [...] These are considerations, but in the end, they are calculated, of course. Often enough, it works. Or you only invest, low-risk-wise, when success is already apparent, right? I mean, you can use many strategies here. *(Interview from 30/04/2021, Karwen, Private Investor & Innovator, own translation of the German transcript)*

Karwen openly says that telling untruths or concealing factual findings is common to convince others of one's idea. The concealment has a double function. First, it protects one's motivation and helps avoid being demotivated by potential repercussions. The narrative thus has an affirmative function that manifests itself later. On another occasion, Karwen tells me that belief in oneself is essential to maintaining an idea's ideals and keep trying before giving up prematurely. He talks about how some development processes can be particularly tough over the years, also because there is not always a market or there are other hurdles in specific contexts, such as funding or legal clauses that are country-specific, as well as directives, such as data protection regulations or even social restrictions, such as pandemics, that suddenly change needs completely. In addition, concealment helps convince others who should also believe in the idea, the product, or the company so they co-finance it. A reliable relationship with investors is fundamental.

Karwen is relaxed about the additional problem of potential fraud. Even in his occasional role as an investor, he speaks of calculated success and failure. He says that to be able to check whether an idea is an idea that possibly promises more than it can deliver, one would have to invest much time. In this context, he equates time with money, so he does not necessarily see a difference here, even though he values well-informed decisions. Nevertheless, he believes that there are different investment strategies. Ultimately, it depends on how much profit one wants from the idea, as the higher the profit, the greater the risk.

Interestingly, he later mentions that these things have to do with experience. He talks about developing a feeling for when an investment is worthwhile, apart from the fact that people 'in the scene' also know each other to some extent so that it is easier to estimate to what extent something is worthwhile. By intuition, he means a feeling that one relies on inwardly and that thus represents a compass or navigator in these decisions. According to Karwen's description, it is a kind of learned inherent knowledge, in both a positive and negative sense, which is emotion-based. However, the problem of the unverifiability of an idea remains, whereby inherent knowledge does not help to overcome the final insecurity in this context. Nevertheless, the concept of knowledge can remain as such; ultimately, it is about a form of knowledge that is not quantitative but qualitative-situational and, in this respect, has a meaningful value for the person who has recourse to it.

For all cases presented, acting pretentiously is part of the business when introducing one's idea. 'Fake it till you make it' becomes a necessary belief and a strategy to overcome and negotiate the hurdles of insecurity. This phrase is part of an entrepreneurial culture I have encountered in incubators and makerspaces. Not only is it ubiquitous, but it is also inhaled and incorporated. It not only blurs one's insecurities, but it also makes faking facts acceptable to a limited extent. The justification for it lies in the culture of mischief that gave rise to the saying and which it reproduces.

VII. Emotions as Valuta

This final empirical chapter delves into a significant aspect of innovation- the role of emotions. It explores how emotions are not merely a part of interpersonal product developments within and outside the team but are also managed as a form of currency. This novel perspective assigns emotions a moral value that can be commercialised and capitalised on, thereby shaping the innovation process in unique ways.

This section thus identifies emotions as commodities and asserts that capitalist logic has shaped a distinct emotional culture (subchapter 4.1.2.), particularly regarding the initial aspect of IP—namely, the legal ownership and interpretation of ideas, reflected in the diminishing of ideals throughout development. Once an idea has created a goal, it is a matter of commercialising it. It is vital for the parties involved to clarify questions about IP and shareholdings. These situations are conflictual, emotional, and often justiciable.

In the further course of this process, the importance of knowledge about a future product becomes visible. How a product is conceptualised and narrated and how these dreams and narratives are further developed points to communication systems with a social function (see subchapter 4.2.), be it a corporate identity or a specific habitus that develops within a team. As discussed earlier, these values develop within a group and promote reciprocal expectations in social interaction. It becomes a social system that is reflected in the prototype whereby both the corporate culture and the communication concerning a product are emotionally managed. Ultimately, so-called *demo days*, a term commonly used in the accelerator industry, convey this through a show where teams present their achievements. Special tools and drama lessons help teams to systematically prepare to present themselves and their prototype as effectively as possible. Such demo days are where markets arise, commercialising the belief in oneself and the product. Ultimately, this partly reflects the strategies of innovators who know how to reify emotions. Barometers are a measuring method at these events: the applause or a vote selects the best development. Demo days are usually exclusive events, sometimes held in an intimate atmosphere, that only invited guests are allowed to attend.

Accordingly, in this chapter, we find different forms of appropriation preceded by an evaluation of the artefacts: claims of partnership, communication systems of (self-) realisation, and performance methods.

7.1 Claims and Ownership

As Chapter VI described, some potential conflicts and flashpoints can occur during the innovation process and within teams. In this chapter, however, a different type of conflict is introduced, which can also be of an interpersonal nature, although this is of secondary importance in the analysis. In connection with a consideration of emotions, which is dependent on commercial purposes, we encounter, above all, insecurity, perplexity, and sometimes fear. Why are these emotions related to commercial purposes? In my observations and interviews, one subject emerged prominently: disputes over funding, intellectual property, and patents trigger emotions. These conflicts arise later in the development process and become pertinent once the value of an idea becomes apparent, rather than remaining a vague notion, precisely because these issues are not (or cannot be) resolved in advance. With the significance of a technical device, the questions of participation and IP increasingly come to the fore as the ownership shares determine who gains power over the product and, thereby, decision-making power. However, the appropriation of a thing already counts as a value judgement, and the mere expectation that something will bear fruit produces a property interest.

Viktor, for example, tells me that he finds the discussions around equity particularly exhausting. He thinks that founding a start-up would bring enough uncertainty. The discussions about equity, i.e. the question of how much one would financially benefit from one's idea, are frequent and very stressful for many.

V: And especially with, let's say, equity or ownership. Like, that's a big issue. And it was always an issue with these accelerators. Like, you know, what's the percentage they should get? If you go into that, which I don't recommend, it's a can of worms, you know. It's the most difficult topic that every start-up here doesn't want to discuss. [...] And it [the question of equity] also provides uncertainty. Because I was in different accelerators, I know how much uncertainty it is and how many problems people have with this, as in emotionally. I think it's a huge burden. So, like from energy, you're probably a creative person or a young person full of energy. You want to do stuff. And when you have no idea how much equity you would get out of this work because the funding situation in equity (I: Exactly.) is uncertain, this is a huge burden and energy.

I: Does it drain you?

V: Yeah. And you don't want that because making a start-up is already hard, and then having an uncertain environment is harder. *(Interview from 14/06/2020, Viktor, Developer at Health Hub)*

For Viktor, in particular, the situation is unusual because he has an employment relationship with the incubator, which means he is an external contributor. In the meantime, his team has disbanded, and he is the last remaining member of the original team. He cannot be a team leader due to the incubator's regular structure, even though his original idea was very similar, which is why he was hired in the first place. He brought the necessary know-how as a developer; he was independent and was already familiar with the vision. Now, however, the incubator is primarily available to doctors at a particular clinic, so he only has a temporary employment relationship. Although he invests his ideas, knowledge, and time in the incubator – and at present almost exclusively – he has no claim to share the finished product as the rights are the incubator's prerogative. Viktor speaks openly about his insecurity, which he accepts in exchange for continuing to work on the idea. To what extent does he build a plan B on this? He only lets on to a limited extent, although he seems to have some plans. Although he is not pleased with the situation, the incubator does seem to be a stepping stone for Viktor.

Nevertheless, he is aware of the precarious situation, which entails a great deal of uncertainty for him. He had hoped for more, especially in financial terms. In his current situation, he feels functionalised and not valued. He is aware of the discrepancy between his work performance and the share assets that do not exist for him. This discrepancy results in disdain for him as a person and for his output precisely because he is the only remaining team member—apart from the original head doctor, Bahar's mentor—who has continued to monitor the project's progress.

Ryan, who works in the same incubator, describes a similar conflict, although he is still in the early stages of his product development. In his work, the conflict becomes even more concrete and justiciable. The technical development team, a company that receives orders from Ryan to technically realise the idea, is paid with funds from the incubator and claims ownership itself. The conflict emerged through the former collaboration of a chief physician with the tech company, who was initially involved in developing the idea before he transferred it to Ryan. Hence, the company was already involved when Ryan was yet to be accepted into the incubator with this idea. The problem could not be solved so far, and previous talks have yet to lead to a solution. Ryan feels pressured and is afraid that the project will fail.

[Health Hub] has said, "don't do publications until the IP is secured. Don't publish anything. Don't say anything". And so, my boss would, my boss is, he's a professor. He lives on publications. He wants to publish stuff. The [incubator] said, "rather not, don't do it until there's a kind of commercialisation or IP". So, again, this is also another conflict with my boss [...]. And I think my boss is also a bit under the circumstance or the idea that this is also just a big research project as well, which it is not. It's more of a commercialisation project. *(Interview from 04/12/2021, Ryan, Physician & Innovator at Health Hub)*

Furthermore, Ryan emphasises a significant challenge stemming from this conflict: he is presently unable to publish papers on the idea or data until the IP issue is resolved. In the course of this, he explains that he and his supervisor would like to do this after all because they see it as part of their medical research profession. As before, he refers to the *Theranos* problem, which suffered, among other things, from the fact that no data were published. However, both Ryan and his mentor are urged to remain silent because the conflict with the other company is unresolved.

Consequently, Ryan is not only unable to contribute to the scientific discourse, but he also increasingly feels under pressure. Not only does he feel like a team lead, as he tells me later, but he is sometimes involved for days at a time in playing a mediating role between the two parties, namely between the incubator and the external tech company. This wastes time that he could otherwise use to progress in his work and with the team. Further, it disrupts his work and causes him to feel pressurised. The resulting feeling of irritation relates to an uncertain work terrain in which he does not know whom to confide in, where to disclose and where not, and where the pitfalls lie. He works cautiously and often feels demotivated as a result. In general, another problem becomes visible: the incubator, which is supposed to create security through its integration and advisory services, becomes an insecure employer due to its set-up, primarily because public funds finance it. Not only is the incubator accountable for what happens to the money and what the incubator and its teams use it for, but there is also the problem that it cannot act freely in business connections: whether equity or IP, the incubator wants to refrain from entering any cooperation, which it would have to justify.

In a conversation with the head of the incubator, it also turns out that the incubator did not originally intend to fund vague ideas but only advanced ones. However, this could not be enforced in the past, and thus particularly immature ideas were also supported. The situation with Ryan's idea is, therefore, ambivalent. If it had been very immature at the time of the application, there might not have been any conflicts with the tech company, but on the other hand, the incubator would have had to promote the idea for much longer than it had intended.

Consequently, insecurity in the workplace and conflicting situations in which loyalty ties dissolve can lead to misjudgement of development factors that can put entire projects at risk. At the same time, this increases the susceptibility to manipulation. If team members no longer feel committed to the incubator and loyalties dissolve, the consequences can be far more extensive. The supposed rationalisation of this process, i.e. looking at a product without understanding interpersonal relationships, their communication as well as emotional aspects, results in the alienation of the team from its product. It is not only the artificial isolation of the artefact from its origins but a subsequent privatisation that becomes the opposite of a preceding creative process as an existential activity. The product is alienated and is no longer an idea in the original sense. The idea belongs to the innovator, which they take to the outside world. Even if the idea is inspired by the outside and triggered by experiences that take place consciously, as Dewey described, it still has a human creative origin that is de-idealised in the expropriation. In this way, the innovation process relinquishes a potential that it could otherwise harness. The awareness of the problem that precedes the creative process means the closeness and empathy of people with each other. This intimacy is given up through the constant process of reduction through the product's commercialisation. The artefact could be an expression of interpersonal understanding, but in this process, it does not remain so.

7.2 Evaluations: From Self-Fulfilment to Gilded Futures

In the context of innovating and prototyping, evaluation processes are constantly at stake. The prototype itself is reviewed, goes through test phases (subchapter 4.1.3.) and is adapted correspondingly to its actors' actions. Similarly, this happens within a team, the incubator, or with external business angels. The team, it seems, develops parallel to the prototype, evaluates itself, argues, and re-evaluates itself, and the incubator does the same as it evaluates and advises the team (subchapter 4.3.1.). However, there is a clear difference as the financier is dominant, and everything that serves commercialisation purposes has special significance in evaluation processes. Therefore, the question is to what extent incubators or funders see themselves as part of the formed moral economy of a team. Related to this is the focus of the evaluation logic for a product. Ultimately, a defining factor is the evaluation logic applied both in the course of development and especially at the end of development. This phase corresponds to the last opening of an iteration in which (perhaps) final changes are made or a narrative is adapted.

Incubators now follow market logic, which in medical production must meet specific standards, such as the *Conformité Européenne* (CE) seals or the US *Food and Drug Association* (FDA) regulations, for accessing the US market. What is also concerning is the often-lacking perspective of potential users, as suggested with CTA, which occasionally matters but is often disregarded.

In this subchapter, I focus on my interviewees' perceptions of incubators' or financiers' assessment practices. Therefore, I look at the perspective of incubators or external advisors who are not part of a team but evaluate the products.

First, Felix, the incubator's externally contracted consultant, discusses how the incubator decides whether a project is eligible for its assessments. He compares this process with the evaluation processes for students' advancement to the next level of their education.

> The accelerator's main task is not the demo day but to filter out which teams are still worth funding. And I always find it like school. Who gets into the next class or is allowed to take the Abitur [A-levels] or something? And that is determined by the ability of the students. And, of course, also on the potential of the product that someone builds. Is it a world-changing product? Then, of course, you can put a lot of effort into it. Not only about the student. And this further support – I think it's called phase 2 support. This follow-up support. There are teams that in turn have funding and need further support. But it is entirely up to the team to decide whether they want to work with me or not. [...] I have to sell myself or my services.
> (Interview from 13/07/2020, Felix, Consultant at Health Hub, own translation of the German transcript)

He describes different stages fundamental in this incubator, with the initial challenge revolving around the successful application, followed by a subsequent phase determining the extent of additional funding for the project beyond the initial project period. In this context, capitalist evaluation criteria play a significant role, which Felix describes with the question: 'Is it a world-changing product?' In concrete terms, this means that the team's previous milestone plan must be adhered to, an initial market analysis must be available, initial data must be available to confirm the market analysis, and the incubator must ultimately see an opportunity to find a buyer for this product in its network. However, as an external consultant, Felix also has his own interests to represent, and he is subject to the team's evaluation processes, as he has already revealed in the excerpt. Even if the incubator initially hires him for a team and is available for it, he says he has to 'sell himself' and his service. He is also subject to performance pressure to have his work and its value recognised. If a team does not want to accept his work later, as Bahar's team decided, he obtains fewer contracts from the incubator. Accordingly, he, too, is interested in promoting the projects, especially because they value him and his work and regard him as an asset.

The interests of the incubator go far beyond the successful out-licensing of the later products. Jan, the head of the incubator, tells me which overriding interests matter: he also focuses on the out-licensing of projects; however, this aspect is new and one of many essential points for a profitable business. Because public funds finance the incubator, it adheres to other constraints and bureaucratic structures. Unlike any private investor, the incubator is accountable. Moreover, it is a pilot project of the federal government and a German federal state that has to be worthwhile. The project, which has received funding amounting to several million euros in recent years, is not only obliged to keep expenditures transparent but is also accountable for expenditures and losses.

> This means that we are now increasingly attracting the interest of the [federal] state (..) when new companies are established here that create jobs with the ambition of paying taxes, which is good for the location. So, now we are at the interface of technology transfer or translational medicine "digital" – home of the digital lab. [...] These are regularly presented to external juries, external bodies that can cover medical aspects, clinical, technological, and so on and so forth to assess which projects are more promising or which should be funded rather than others. This is how we select the projects. *(Interview from 13/08/2020, Jan, Head of the Accelerator Programme at Health Hub, own translation of the German transcript)*

Accordingly, as Jan states in this excerpt, the accelerator programme must be worthwhile to guarantee continued funding and convince the federal and state governments, alongside a parallel private excellence initiative, of their value. If the investment is worthwhile and the incubator can shine with successful out-licensing, it will continue attracting doctors who want to realise their ideas. The incubator thus becomes a flagship incubator that uses public funds to promote the biomedical field.

Figure 15: Jan's Perspective: From the Idea to Selling the Product

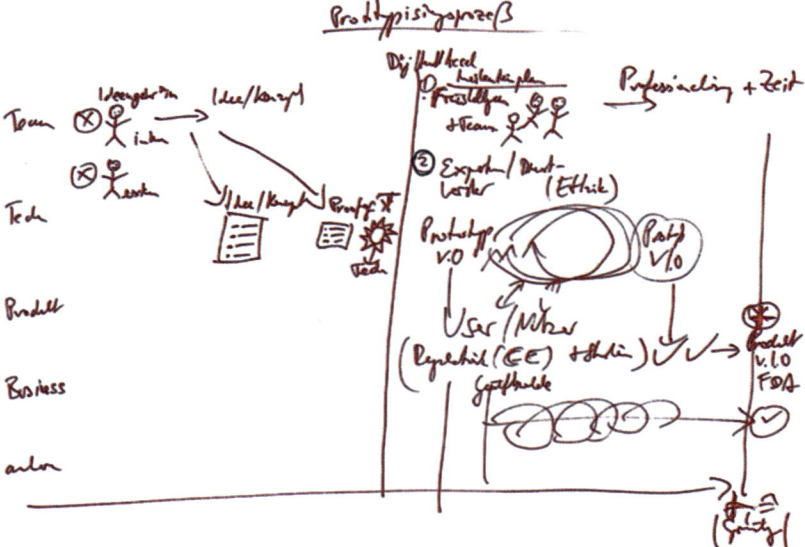

Thus, the justification pattern revolves around the potential success of out-licensing or the emergence of companies from the projects. These companies then establish themselves in the region and contribute taxes to refinance the public money invested. To ensure the success of the ideas and projects, external juries are invited to the aforementioned demo days to give a guiding opinion that should help them succeed. The jury members are people from the public sector with a professional and related background, health insurance companies, and other experts from the health sector. The jury members change with each demo day.

Felix allows an insight into one of his projects with regard to the work processes:

> At the moment, the focus of our work is that we are developing a piece of software and are currently helping the team to get funding by working out the necessary paperwork and launching the first version of the product that they are talking about as a test version. That's happening next week. And this test product will then be run with various doctors and therapists – it's a therapeutic, physiotherapeutic product – as a so-called closed beta, i.e. for a closed circle of users, in order to be able to recognise whether it will be accepted. On the one hand, by patients, on the other hand, by therapists and also accompanying specialists. To use the numbers that are generated in the context of fundraising. By being able to say, hopefully, that this is encouragement, or we call it, traction. And one can use that as an argument in fundraising. That's what I'm doing as a focus at the moment, with a team. *(Interview from 13/07/2020, Felix, Consultant at Health Hub, own translation of the German transcript)*

This insight is the first time patients have been included in a test, although it is only a handful. 'Closed beta' is the process name Felix describes above, testing the prototypes. It is a test scenario in which different actors come together with their expertise and now communicate with each other around the artefact and evaluate it together. These processes were previously impossible for reasons of patient and data protection. In this evaluation process, the product undergoes assessment from various perspectives. However, the patient evaluation is only gradually included at this point because the end consumer is a specific one, and it is a medical device. If this were not the case, an end consumer would be invented.

Nevertheless, the invention in question here aims to address a specific medical condition and is assigned a unique role in this context. Accordingly, suffering is an independent experience that any other person cannot otherwise comprehend; it is itself an intentional phenomenon as it is a personal, subjective state that generates its own form of knowledge and truth. The state itself is thereby autonomously emotional. If more space were allowed for the patient's perspective in general in this process, as the CTA concept recommends and as I highlighted in subchapter 5.4., a different dynamic would emerge that would generate a far greater empathy awareness. If one follows a philosophical view, suffering is a denied state of will, thereby describing a state of inadequacy. At the same time, this state can cause completely different states of activity to eliminate or alleviate the state. Ultimately, however, the discovery of the problem is insufficient to fully exploit the creative process. Hence, if the person who lives with a problem would be involved in solving it, far more could be accomplished. The problem of a lacking or diminished user perspective is considered again in the following subchapter.

7.3 Demo Day: Performing Emotion

The reduction processes continue. In the previous subchapter, I looked at the perspective of the incubator and external advisors. Therefore, in this section, I focus on the teams themselves. Just as the reduction processes occur in developmental de-idealisation, we reencounter them when communicating the prototype. The dream is initially idealised and is also later functionalised for practical reasons. In this subchapter, I deal with the demo days mentioned previously, during which the ideas become a projection surface for more *purposes*.

At the same time, the presentation on the demo day serves as self-realisation, which then gains social recognition with the term *purpose*. Whether in Berlin or other major cities, terms and phrases such as 'self-fulfilment' and 'make your dream come true' characterise the start-up scene. Unlike in Chapter V, it is not about the initial imagination that already draws a picture of the future but rather concerns changing narratives that are adapted. These narratives are directed at developers and the product itself to establish a purpose while simultaneously adapting this purpose to an existing or emerging market.

In this context, it is no longer about what the innovators and developers feel in the production process but much more about how a prototype, isolated from its original idea, can be developed into something that can generate a market and attract consumers. In this respect, it is not about the narrative of self-realisation but the narrative of purpose realisation. This ambivalence, in turn, does not follow the emotional self-expression of those from whom the idea springs but is about creating a sense of well-being in those who will ultimately encounter the final product. The focus thus lies on creating a feeling of well-being that results from the artefact and elicits it in the consumer. However, this feeling of well-being is not judged by the consumers themselves but evaluated on their behalf. Reconsidering the sketch in subchapter 4.3.1, we observe the last step of parallel development, in which new value development standards are established. As shown in this subchapter, the now emerging parameters arise in the context of final iterations.

Figure 16: In Preparation for the Demo Day during the COVID-19 Pandemic

Figure 16 shows the preparation for such a demo day, which took place online, so participants and technicians checked the necessary equipment when I took the photo. A small collective from the M.lab organised this demo day, and after much effort and overcoming regulatory hurdles, they found a space and let the long-planned demo day take place. Some had been waiting so long to present their projects that they were worried about finding a purchaser. Especially during the COVID-19 pandemic, financing phases ended, and some lived on their savings, as access to interested parties and markets was difficult.

During my empirical study, I visited several demo days, both small and intimate ones, as portrayed in Figure 16, both in person and online, and bigger ones organised by giant tech companies. The atmosphere is tense and exciting, especially on the bigger demo days. People who, I believe, are developers, designers, and innovators run around, looking for people to explain something to them at the last minute and greeting one another. Microphones are installed just before stage appearances, full headlights are turned on, a team lead is introduced, and the product is presented.

Bahar tells me about her experience on the first demo day at the Health Hub incubator. The incubator rented a particular location for this event; there was a stage and technical equipment such as microphones, large screens, and cameras. She talks about her nervousness, which she also explicitly trained with a drama teacher beforehand. After the presentation of the prototype and the idea behind it, many of the jury members and guests approached her.

> There was a special day when we had to present everything we had developed. And then everyone approached us. The health insurance companies were interested. And there might have been funding and so on. The first LOIs [Letters of Intention] came in, and so on. *(Interview from 30/01/2020, Bahar, Physician & Innovator at Health Hub, own translation of the German transcript)*

She tells me in conversation that she was asked many questions that displayed a particular interest and occasional hints on what else to look out for in the development were provided. She briefly summarises the interests of the individual actors:

> This is a microcosm here. Everyone wants to get something specific—me, a product that corresponds to my idea. [The incubator] wants to know that we are developing something worthy of being funded. The insurance companies want a safe product, which is why we have the Johner Institute on our side [...]. *(Interview from 30/01/2020, Bahar, Physician & Innovator at Health Hub, own translation of the German transcript)*

A good impression must especially be made in front of potential customers, i.e. health insurers, and it must be possible to assure them that the product is certifiable. Concerning the latter aspect, there is a German institute, Johner, which takes care of the fulfilment of the criteria. As another external consultancy, it advises projects such as Bahar's in the med-tech sector and is familiar with the regulations for medical technology. It helps the incubator and the teams review the safety criteria and the quality segment up to the product's approval.

Bahar continues to talk about the dilemma she faces in reconciling the different interests of the patrons.

> You always have a goal where you see, okay, you've completed something again, and what you've done is rewarded because you get funding again. And then everything between these periods is what you have to do with the funding, that's

always difficult. [...] Especially with the milestone plans, you often think that the patients are actually missing out. *(Interview from 30/01/2020, Bahar, Physician & Innovator at Health Hub, own translation of the German transcript)*

On the one hand, she is pleased about the recognition and financial resources that every presentation success brings her. On the other hand, she says she feels tremendous pressure to perform, and she finds it challenging to work toward meeting the expectations of others. It is a marathon, as Viktor also described earlier. As Bahar herself said, the actors have different interests. On the demo day, the incubator expects her to deliver a performance that arouses so much interest among the guests, ultimately leading to buyers for her idea. Her idea and her original interest in producing an insole supporting patients after knee or hip surgery is only in the foreground as the development context is now reduced to a narrative she can perform.

Nevertheless, she notes that the patients are not the ones who are actively involved in the development process, and the presentation on the demo day features the product in isolation as the patients are not present and play no role in it. In the spotlight: an advertised product without being demonstrated to its intended user, the patient, through whom the problem becomes apparent in the first place. However, if patients are not at play, they are theorised by the isolation, and the once practical reference recedes into the background.

Ryan tells me how the preparation for demo day works and mentions that many people are very nervous beforehand and that the pressure to perform is exceptionally high on this day. One wants to deliver a good presentation oneself, and the incubator also wants you to deliver a respectable presentation. Thus, there are lessons and exercises to help individuals to say the right thing at the right moment and to tell a story, preferably an emotional one. In this excerpt, Ryan shares what the workshop and training were like:

R: She [drama teacher] gave us before we were supposed to do this pitch, this presentation, she all gave us kind of a seminar in how to present, which is great. I had a lot of fun doing that because you're not just supposed to stand up there and just click through a bunch of slides and say, okay, I'm X going to explain this publication. What's more, I'm trying to sell something. I want to deliver emotions to something. [...] How do you connect with these people? How do you look into someone's eyes and say, okay, this is the best thing ever? And here's, it's going to change everything. That was really good. And that gave me a lot of confidence when we had to do these pitches because I was not nervous at all. [...]
I: So, what would you say is what is it about? So, you said confidence is something you'll be trained in or anything else?
R: The clarity of your message and not getting caught up in the scientific details of something, but really just on an emotional basis. What is it you're trying to do? [...] You're going to be presenting this. There's going to be cameras there. There's going

to be lights there. You don't want to freeze. You want to be able to think on your feet and kind of just flow and that. And I think that was [...] the part of the workshop that really stood out to me. *(Interview from 04/12/2021, Ryan, Physician & Innovator at Health Hub)*

Ryan clearly says that it is ultimately about delivering; it is a performance in which one is supposed to tell a story that triggers emotion. It is strategic storytelling, how you tell it, and what words you choose besides body language. The scientific facts are irrelevant at that moment – Ryan was not presenting the actual inability of the prototype nor the IP problems. You want to convince, despite all the nervousness. The cameras, the lights, the many people you have to address. You are given a strategy to overcome all that and stay in the flow. It is the story that counts.

The field notes in my diary, as well as the interview excerpts from Bahar and Ryan, show how performers have to functionalise emotions on the demo day. This insight is well-known, as we have understood advertising for decades. What is more intriguing is the recognition that emotions are assigned a value that cannot be replicated through other means. Emotions serve as mediators of value and must align with the zeitgeist and the audience. In addition, the story around the prototype must make the problem it solves tangible. In this context, problems are easy to emotionalise. The performers must, therefore, recognise and address desires.

7.4 Emotions as a Product

Apart from the aspect of performance, i.e. the way something is conveyed, the content comes into focus. According to this, a potential customer is not only convinced by one's appearance but also by the narrative itself, the chosen words, and how the narrative is symbolically and emotionally loaded. It is not only about 'performing' but also about a speech act, 'performing a narrative', which ultimately becomes a 'performing emotion'. Physical performance, speech act, and symbolic wordplay become one, blurring into each other. In this narrative, however, not only is a potential increase in turnover or sales value expressed, but the approach to the artefact. First, such narratives express a creative force charged by one's experience, goal, and purpose. It becomes an object of knowledge (Dickel, 2017; Knorr-Cetina, 1997), successively enriched by multiple perspectives, by further experiential knowledge added to the object. A reduction process can occur parallel during this accumulation of knowledge, expressed in the artefact with its changes. As shown in the previous sections, the level of idealisation is reduced, and functionalisation occurs.

In subchapter 3.3, I described the 'rise' of emotions and renewed attention to them through (neo-)pragmatism, which resulted in their increased importance since the 1990s. Moreover, emotions have been given a great deal of space at various levels in the social sciences, philosophy, and psychology. They are perceived and analysed in knowledge accumulation processes, considered in negotiating a moral economy and are part of the agreement, whereby the moral economy helps a group to be its reference system on which it agrees.

In the process of reduction, however, emotions change their meaning as, although they are still world-making, they have become flexible in marketing a product. While they used to be significant in communication within a team around an artefact, they are now functional companions.

The following three interview excerpts show how a narrative changes to reach more people because its former content is unsuitable for a wider audience.

On the demo days and in the conversations, it becomes clear that the pressure from incubators or financiers can be tremendous if an idea fails to meet the desired criteria. This pressure can, as described above, move developments in predefined directions that the teams do not intend but ultimately follow to continue receiving funding.

During our last conversation, I asked Karwen about his familiarity with these experiences and sought confirmation regarding the necessity of adapting and imbuing narratives with different significance to ensure a continuous flow of funds.

There was once an invention – but it wasn't mine – about a sensor, haptic models. I got involved as an investor, and later as a partner. Why? […] [I come from a country where there is corruption and war. I have always said to myself, nothing that has to do with these things. No war!] This sensor thing – it doesn't exist anymore like that […] (laughs) – we wanted to use it in the rehab, health sector, I won't say exactly what, because – it's too delicate now, ok? (I: Sure.) We needed more money from a certain point on. So, pitch after pitch and so on. So, the OEMs [Original Equipment Manufacturers] came, the car manufacturers. And then, of course, it was somewhere more obvious to say, okay, we're also going in this direction. Although we still wouldn't have pushed it so hard. Because it was not an emotional, important topic for us. […] And then it was like, hmm, I don't know if we want to go into the car industry now. […] That they said: OK, they think it's great, they see the potential there. We should somehow continue to work on it. And that has also influenced us, but of course, it also comes together a bit with the fact that we have also, so to speak, rationally considered for the company, what else can we do, how does it make sense? […] And so, I left later. (I: Why?) I'll tell you why. The thing was, as soon as it landed at Bosch or a larger car company, and that went quickly – amazing when it comes to money – but, somehow, I got worried. I thought, if they get this, then it can also be used in tanks, on rifles etc. I didn't want that. *(Interview from 08/02/2022, Karwen, Private Investor & Innovator)*

He tells me about a former invention, haptic technology, which he and the developer wanted to use in psychiatric rehabilitation medicine. Although he had already co-founded, new investors were needed as they still did not raise enough money to push the idea further. Subsequently, they came into contact with so-called OEMs and presented their idea to them. The diversions via the automotive industry were not planned this way but made it possible for Karwen and his partner to continue with their business. Later, he says, he became sceptical and dropped out. Because of his background and his experiences in his home country, he was worried that the idea would be repurposed and implemented as war technology. He could not reconcile himself with that and is no longer part of the company. The invention has taken a different path than initially planned as a prominent vehicle manufacturing company bought the idea and its related technology.

Bahar tells me about her invention in adapting narratives, which has had the same name from the beginning, although the explanation of its origin is now different. The fact that the topic of war matters to both Bahar and Karwen – as can be read in the excerpt below – is a coincidence. However, it is interesting that both come from countries that have suffered or are suffering from (civil) wars. The connotation of war evokes negative feelings in both of them, and thus, they do not want a war associated with their invention or for it to potentially be misused for this purpose through application in the armaments industry. Bahar tells me that the name of the project and the prototype that is being created originally came from her mother tongue and the Latin word for foot. With her sister, who developed the first prototype with her in the living room at the beginning, there is a lot of personal and emotional content here. It is not only an emotional matter for Bahar, as she initially describes, to develop such a product for her patients. It is also a private matter with her sister, who includes part of her identity in the name, in the object itself. Together, they stand in their pyjamas in the living room, as Bahar tells me, tinkering with their prototype and giving it a name that – for them – has an identity character.

> At the time, my sister and I chose [product name¹], and [x1] means [x2] in Kurdish. We googled it back and forth a bit and came up with a pun on foot. [x3 is x2] in Latin. And then we had a Kurdish-Latin word. In the meantime, we have changed it for the public. Because Kurdish always sounds like a national war. Accordingly, we now say that [x1] comes from foot linguistics and has analytical foot design. We understand the language of the feet, that's how we frame it now (laughs).
> *(Interview from 30/01/2020, Bahar, Physician & Innovator at Health Hub, own translation of the German transcript)*

---

1   Words are left out for confidentiality purposes.

Bahar and her sister, who dropped out of the course of the project due to her medical studies, superficially give up this identity-forming part for the public, as Bahar says. A Kurdish word, she thinks, would not sell well in the public perception. It sounds, as she says, like a national war and thus cannot generate an outlet or high sales figures – at least, that is her assumption. Even if the identity-creating character is only superficially abandoned, it has a clear objective as the name remains, but the rationale in the narrative is adjusted.

In the interview with Ryan, things become more concrete about the commercial aspects, whereby he first talks about the reduction processes as described above:

> [...] The narrative completely shifted, completely. (I: Can you tell me how this shifted?) So, again, with this original idea, it was just the company, my boss and I are like, hey, wouldn't it be cool to have this? And it's probably going to help patients after surgery; it's going to help. It's just going to help, it just broad, like, oh, [...] we can monitor brain perfusion. Finally, it's just going to help everyone. The narrative shifted during the first phase of this accelerator program, where we had intensive mentoring where people ask like, "Okay, we're not going to fund something that sounds good. Will people buy this? [...] Will hospitals want to actually give you money for this? How can you actually integrate this into a healthcare system?" You can't just say, oh, this is a cool thing. It has to have some value to it. And going through these talks of, again, this is this wonderful thing about this program where you take a doctor, and then all of a sudden, they're confronted for the first time with building a narrative, building a motto, building a business, having more of a business side of view. This really starts putting a lot of reality and roadblocks in front of you. And so, the narrative shifted from wouldn't it be cool to have this, it would be the best thing ever to really saying, okay, what does it actually do? Why would you want to pay for this? What's the story we're trying to give this? What's the use case scenario? And so, the narrative has been a lot more refined now, as opposed to saying, oh, this is going to help elderly people.
> So, there is no data on delirium after operations, too, saying we can actually pinpoint possibly where the profusion deficit lies. So, the anaesthesiologist can intervene and, therefore, outcomes or hemodynamic outcomes for all patients will be improved. And so, that's kind of the story. And so, we say, okay, for the first time ever, the anaesthesiologist can measure this for the first time ever. It's going to be a breakthrough in everything, and this is why a hospital should buy it. So, again, my boss never had this idea, I've just been doing the work thing. Okay. [...] How can you convince someone to spend money on this? *(Interview from 04/12/2021, Ryan, Physician & Innovator at Health Hub)*

As Ryan has described before, the incubator has a great deal of influence on the development, which Ryan perceives as putting him under pressure at times. Furthermore, he outlines how much the narrative has been adapted throughout progress and what questions the incubator asks to guide this process. First, it is the enthusiasm that he, his supervisor, and the developing technical company feel for the intentional idea. In our conversations, Ryan always emphasises how great he thinks the idea was at the beginning and how excited he is to build something, research, and create something with his hands. This is where the aforementioned will to create is active, primarily of an idealistic nature. The solution to a problem and helping the patients is in the foreground. Although the plan sounds good, it is not enough. With Ryan's successful application to the incubator, hurdles are now coming his way. A product whose idea sounds good is replaced by capitalist market logic, expressed in the following questions: 'Will people buy this? Will hospitals want to give you money for this? How can you integrate this into a healthcare system?' The inquiries are now concretising the objectives, and consequently, there is a narrative specification.

The rationalisation process turns the object of knowledge and emotion into an instrument. The outer shell, the new narrative, remains emotionalised to sell it. However, in terms of content, it diverges from the original intention. Rationalisation leads to a return from the emotion object, which becomes the knowledge object, to an archaic stage of being a thing (Knorr-Cetina, 1997). Ultimately, this reduction and rationalisation serve the purpose of raising external expectations. Presumably, it is reloaded in further dealings, in the sense of a continuing sociality with the object, and thus gains new meaning in a new context. However, this transfer of knowledge and emotion needs exploration in further study.

Ryan's presentation on the demo day also provides an intriguing insight. Whereas during the first minute, he talks about avoidable patient harm during anaesthetic operations, the second minute focuses on the costs that potentially harmed patients incur due to anaesthesia during an operation. Overall, 10% suffer perioperative complications that lead to prolonged hospital stays and Intensive Care Unit (ICU) care. Ryan continues with facts and figures: 45 million operations took place in the EU in 2019, 5 million of which experienced severe perioperative complications, resulting in €200 billion in extra healthcare costs.

It is striking that despite their fact-rich description, the narratives cannot avoid underlying the facts – as far as they are such – with emotions. The idea, the motif, and the associated emotions economise during the reduction process. Whether in the incubator or supported by other business angels, the 'idealist' has to surrender to capitalist economic logic in the end.

The ongoing reduction processes, as elucidated earlier, redefine the value of the prototype, departing from the initial values attributed by the inventor, innovator, or team. The standards differ in the course of the iterations or by their end. As we have already seen with the demo day, it is about an emotional performance that grips the audience. The product narratives function similarly, as they are not solely performed, yet they must convey the content effectively to move the audience. Yet, as I described, the narratives or stories are adapted to an audience, their world of experience, and acceptable emotions, which depends on where the narratives occur. Some benefits are emphasised, while others are relegated to the background if highlighting them does not seem justifiable at the time.

We have seen that Bahar adapted the story to her product name as she suspected and felt that a name of Kurdish origin would be less saleable. Possibly because concerns about inherent racism in evaluation patterns concerning her product are not far-fetched, and the Western world would instead source consumerist goods from its own climes? Addressing this question in detail would provide plenty of evidence. In any case, her assumptions and concerns are an expression of socialisation or part of the experiential world of a given group.

Karwen, on the other hand, decided against producing technology suitable for war, among other things. He maintained this ideal for himself, which is why he abandoned his team and the idea. Maintaining his ideals eventually came at a price.

Finally, we encounter Ryan, who also describes the transformation of the narrative, albeit in favour of saleability. As he recounted earlier, narratives are adapted but include the weakness that they did not – at least at times – point out that part of their promise was based on speculation but simply accepted this for popularity.

All three are confronted with the functionalisation of emotions, not necessarily in a moral manner, but to create a want or perhaps satisfy a need, but always with the purpose of being relatable and saleable.

Ultimately, it is remarkable that rationalisation during the development and production of a technical artefact, such as medical technology, is invariably countered by emotionalisation. Ungrateful remains the abandonment of emotions, which are supposed to be excluded during the reduction processes, the ruptures that arise due to rationalisation. However, emotions are only requested when a calculating purpose benefits them.

VIII. The Moral Economy of Different Intentionalities

8.1 From Radicality to Reductions to...

Looking back, I would like to return to the concept of moral economy as we know it from Lorraine Daston, who defines it as 'a network of affect-saturated values that stand and function in well-defined relationship to one another (Daston, 1995: 4).' In this context, I described how the concept of morality enhances actions and objects with emotions (Daston, 1995: 4). I have shown that what is commonly called innovating in makerspaces and incubators are such moral economies. These are places where different people come together to negotiate an idea and a prototype that emerges from it. These places become marketplaces, and the participating actors become market criers, informing about their own ideas, expectations, imaginations, and desires. Studying these actors and their stomping grounds is highly beneficial.

> They answer old questions and pose new ones about how [a group of researchers and surrounding structures] at a given time and place dignify some objects, [in this case, the prototype] at the expense of many others, trust some kinds of evidence, [such as their data,] and reject other sorts, and cultivate certain mental habits, methods of investigation and even character of a distinctive stamp (Daston, 1995: 23).

Accordingly, I have introduced the objects the actors examine and explored the emotions that arise in the context of their creation and evolution. Both the developed prototypes and the directions, as the materialisation of implicit questions, can inform us about the society in which they emerge.

This does not mean the key figures of society give us information about quantifiable data. Instead, the moral economy examines what a group of researchers considers suitable or 'valuable' to devote themselves to, what problems they see in their field of activity, what they disgust or neglect at certain times, what they miss and regret, and whom they trust. We learn about their feelings and, thus, about social structures that give us information about which structures outside the group affect them and how they deal with this, i.e. what obstacles there are and how they overcome them. Their mental habits, methods, and gaze tell us about their culture. Society and groups negotiate emotions. They show which feelings are 'permissible' to feel and which sentiments are socially acceptable. In terms of the triad of *thinking, feeling*, and *acting*, and next to it, the three different mental acts of *presentation, desire*, and *judgement* provide the frame of reference for what we evaluate: how we feel, think, and consequently act concerning something.

In this work, I have shown how ideas arise and to what extent their emergence emotionally charges them. Ideas are thus reflections of what happens to us in everyday life. Within the framework of our emotional and judgemental space, we evaluate what we consciously perceive, absorb it, and use this knowledge to let new things emerge from it. Problems become challenging sources of inspiration in the course of new spaces of possibility, which are supposed to optimise what exists in keeping with the idea of progress.

Accordingly, how we think about and imagine something says a lot about our ontological understanding: how we relate entities to each other to arrive at a reality where we can direct our feelings and values towards something. Ultimately, I have illustrated the many realities related to each other or at least attempt to be associated with each other as soon as an idea or a first prototype is brought into an incubator or makerspace to develop the technology further in a team with further patrons. I want to conclude by recapitulating these numerous realities.

1) The Radicality's Creativity

I found that motivation is emotional for inventors to become active, although it varies from inventor to inventor. I argued that how we feel represents our relationship with the world. *Our feelings and how we react emotionally express our previous experiences.* The ideas from the study arise from the oscillation of the inner and outer worlds. We make new additions to something already there and new combinations from what we know or spontaneous ideas that occur. Empiricism shows us that inventors often regard their ideas as having a saviour-like quality and as a 'Swiss army knife' seeking to improve the status quo. This understanding presupposes that a deficiency has been discovered or uncovered, whereby the perception of the problem depends on one's own experience. It is emotional because it depends on one's view of the world and its reaction. In this perception, there is a world judgement and, thus, an emotion.

The examples mentioned were mainly experiences from everyday work, examined in isolation in selected incubators or makerspaces. Contrary to what is often assumed, these ideas do not arise in a design thinking workshop. The innovators get down to the root of the problems and become *radical*, which ultimately constitutes their intrinsic motivation.

Discovering and collecting the problem is the prerequisite for what can later be called creativity. In the study, *the discovery of the problem is already emotional*. We find frustration, compassion, and burden as emotional expressions of the innovators' everyday observations. Offering a solution, i.e. the answer to the problem experience, is thus the result of what a frame of reference allows. They are, I argue, 'controlled solutions'. *Both emotions and innovations know their limits and adapt to their societies*. This means that not only is an idea managed in the course of its development according to an incubator, team, or milestone plan but also the emotions are adapted to what generally seems appropriate in a cultural framework. However, this phenomenon reaches its peak later, with the commodification of emotions to reach a broader consumer market (see the section on 'Emotions as Commodities').

It also became clear that what the interviewees described as motivation can be emotional. In particular, I think of Bahar in this context, who identified her anger at a grievance as her greatest motivation. However, it is a hurdle, especially in professional and scientific contexts, to name these emotions as such or to state them as a reason. Bahar said disclosing these feelings could be perceived as 'stupid' or pathetic. However, thinking and feeling about the prototype oscillate during the development process. Referring to the control of emotions and the boundaries observed, only the feeling that someone is willing to show becomes visible and collectively negotiated. Here, the first type of reduction begins, which we will encounter more often.

2) The Moral Impact

In these contexts, however, it becomes apparent that there is often a personal connection to the inventor's idea. This observation is not very surprising, considering the emotional content of the reaction to the outside world. My interviewees described a sense of connection, compassion for a grievance, and enthusiasm, excitement, satisfaction, fulfilment, and joy as feelings they experience when they have or develop a solution. A sense of power, as Karwen describes, is given to him when he thinks he can bring about an improvement through his act of creativity.

In this context, the purpose is equally articulated and emotionally linked, swiftly creating the impression that his form of altruism transcends his problem-solving aspirations. I called this state the *moral impact*, meaning *an inventor's activity claims to be meaningful and purposeful*. The described purpose and their conviction find confirmation through narratives that develop parallel to an idea. This confirmation helps to overcome uncertainties during the development stage. In this context, I observed religious parallels as the belief in the idea becomes an ardent desire. However, whether the prototype triggers this fire or the faith in it could be the subject of further investigation.

3) Structures of Innovation

Structures of innovation, such as through and in incubators and makerspaces, *are an expression of a creativity dispositif and the postulate*, be it political or social, *of being and becoming active*. Such spaces allow room for a 'could-be', i.e. possibilities and serve – in the eyes of the innovators – a greater purpose, one that they compose. Such possibilities, i.e. what innovators project into their prototypes, are a materialised expression of utopia. Through the inventors' activities, those spaces come to life. Thus, the structures are part of creative culture and, for this reason, one object of investigation of a moral economy.

Nevertheless, the structures also result in hierarchies influencing innovators, teams, and ideas. In these places, money often plays a role, exerting constraints. Especially during the COVID-19 pandemic, it became clear how many innovators, some of whom financed their ideas with their private funds, got into considerable difficulties that partly affected their livelihood. This problem is more likely to affect those who choose to work freely in makerspaces, as we saw in the M.lab. The situation is different in the incubator, which may pay a salary to innovators if it takes them and their ideas on board – in exchange for rights to the successful product. However, the management of an incubator can then reserve the right to make decisions during iterations or evaluations. The situation is different for Hydro GmbH, which I can only include here to a limited extent. With its breakthrough idea in the 1990s, the company is a well-established and permanent fixture in the field of hydrocephalus valves. This company is a success story in the German innovation landscape and, thus, an established structure. Its established nature means that fewer uncertainties influence its day-to-day business. What is interesting, however, is the observation that precisely because of the security, there are several possibilities to bring user perspectives, i.e. patient perspectives, more in line with the product and, overall, to focus more on taking the patient's needs into account. Accordingly, the problems and emotions mainly relate to the teams that feel more uncertainties due to the external structures.

4) Innovation's Obstacles

During the innovation process, there are continuous hurdles that the innovators, whether individuals or teams, have to overcome. The structural environment in which innovation takes place, who finances the project, and which actors are involved all matter, both in terms of the problem and the solution. As the empirical study clearly showed, *a lingua franca is often the solution or, even better, prophylaxis for potential difficulties*. As it turned out, mutual understanding of the professional backgrounds of the actors involved was a minimum prerequisite for this. It was also mentioned that ethnic background should be considered to avoid misunderstandings.

Furthermore, shared frames of (moral) reference are beneficial. First, it is vital to work out the joint expectations of the project and the goals the team wants to achieve in practical terms. On a meta-level, the emotional frames of reference need to be aligned. This alignment means that common logic and moral concepts that correspond to common maxims must develop. Furthermore, the question arises of whether it is a foreign logic that the team adapts. If the horizons of expectation do not match, this may well mean a criterion for exclusion from further cooperation, be it the structural environment, the financier, or the team.

These questions are also essential since *a strong identification with the project and its work go hand in hand*. The values and logic that initially exist or develop within the team later stem from them as a jointly thinking collective. Making oneself understood is the prerequisite for the collaborative working and thinking process.

As my confidants described, the other side must be understood in its emotional world. *Mutual understanding includes retracing expectations, norms, and values and verifying emotional competence to achieve the mode of sociality*. Otherwise – as I have shown – conflicts arise, as we could observe with Felix and Bahar. Although there are other reasons, both feel they are not taken seriously by the other. Here, too, we encounter the reduction process described earlier, and hence, frustration arises when it is no longer about comprehensibility, and ideals have to be set aside to make more room for feasibility. As described, the market logic then takes hold, creating a different, new pressure, which is then – as we encountered in the incubator – passed on to the teams. I observed an accountability and transparency obligation for using public funds, which depends on success in sustaining its milestone plan.

Interestingly, *a team's efforts to create a common language, common goals, and an identity are then relegated to the background*. Once the incubator activates its role, the language and logic change as marketability comes to the fore, even though the incubator has benefited from the previous team's efforts, which were not necessarily in line with market logic but with the fulfilment of ideals.

As observed in the *Ellie* project, the website presents promises for marketability purposes that are far from fulfilment. Conflicts arise that promote excessive demands and disorientation. In addition, there are misunderstood hierarchical relationships and a lack of recognition or even suppression.

Bahar, for example, adopts a different, brasher behaviour as she feels oppressed and does not want to be ignored. In the decision-making process on prototype developments, we encounter frustration, discouragement, and dissatisfaction, and negative feelings influence decisions and hold further potential for conflict.

5) Trust as a Meta-Emotion in Co-Working Processes

In this context, it is unsurprising that *all the informants working in teams speak of trust as a means of creating a stable environment, which they consider necessary in the vagaries of the entire team and development process*. In the process, two types of trust were encountered, which I categorised as emotive and cognitive, whereby it became clear that the emotive category is often assumed, even if it indicates the other. The distinction would be negligible at first if not for the potential for misunderstandings. For example, when Jan speaks of trust, he means it cognitively and understands it as something a CV could exhibit as an extra skill. A critical aspect that can be ticked off when looking for potential team members is that an incubator has to buy in. Surprisingly, something as interpersonal as trust cannot be dispensed with, despite the logic of the market, although it seems to be a combination of interpersonal sympathy and the need for a market relying on such things. Financiers are aware that no team functions without what takes place on an interpersonal level, as these mechanisms cause people to develop sympathy and trust.

I found that entrepreneurs think of failure as a part of innovating. Apart from the desire to succeed, failure represents a Damocles' sword that hovers over innovators and their projects, whereby, sometimes, failure is a calculated position. The positive reinterpretation of failure is conspicuous in the entrepreneurial scene. When people talk about an 80% failure rate, this high percentage also needs justification, which is why they sum it up in a snappy saying that flits across the corridors: 'Fake it till you make it'. However, this saying also involves simulating results or a success story until they – hopefully – finally materialise. These over-optimistic narratives stabilise during the uncertainty that accompanies teams developing an idea. They are adapted, perhaps a 'tissue of lies', as Bahar describes it and provide justification and the need to implement the idea in times of uncertainty. Entrepreneurs make promises they cannot yet verify, which seems a common practice to maintain funding despite (and because of) best intentions.

6) Emotions as Commodities

Finally, my empirical research showed that *through continuous reduction, emotions eke out an existence as commodities*. An emotion culture develops that knows how feelings need to be managed in corporate culture to develop and become part of the marketing of a product. In particular, questions around authorship and (intellectual) property bring out feelings of irritation, insecurity, and fear. Viktor often describes feeling undervalued and functionalised in his work with the incubator. Ryan, on the other hand, feels pressured to withhold research data until his specific IP issue is resolved. Apart from being an individual problem in this project, this is also a challenge to the development of science, as it becomes apparent that he and his mentor are becoming adjutants for economic purposes and are not allowed to act in their best interests as scientists. They imposed a duty of confidentiality on research data to commodify the idea, which runs the risk of becoming justiciable.

Such imposed regulations and structures in the incubator can be critical from several points of view, although the idea of public funding of innovative ideas is worthy of appreciation. While the accountability imposed on the incubator for transparency reasons due to handling public funds makes sense, conflicts with the internal expectations and self-designed business strategies are shared across the scene. As I had learned in the course of the survey, the incubator management was difficult to reach and accordingly, so were the teams that were placed through the management. Until the demo day, the projects largely remain confidential and only when the prototypes are ready for demonstration will the incubator make promotional films and present teams publicly.

The role of financiers, in general, and the Health Hub incubator remains dominant. Reductions mainly occur during evaluation processes in which the incubator interferes considerably in favour of feasibility. Its accountability – which is not only due to public money – seems to put the management under considerable pressure, which it passes on to the teams. We saw that such pressure manifests in reductions resulting in de-idealisation whereby the idea, with its original conceptions, gives way to a commodification process. 'Closeness' and 'empathy' are feelings that are necessary to get to a problem's root that otherwise fades into the background. With this in mind, the artefact could testify to a successful process of communication and rapprochement, but ultimately does not remain so as feelings of empathy recede for the benefit of feasibility.

During the constant evaluation processes, the question can be asked whether the financier is not also part of the moral economy we encounter here. Moreover, this remains debatable until the end. This argument can probably be ruled out for teams such as *Feety* because the incubator, which finances the development, has long given the impression that it has yet to develop a lingua franca with the developers. This first manifests in the refusal to cooperate with the external consultant and finally in the fact that the original team was falling apart.

In the end, everyone wants to have their share. The incubator wants to be able to sell the rights to insurance companies. The external consultants want to sell their services to the incubator or the teams, and the team wants a functioning prototype that receives a security seal to satisfy the incubator. The incubator remains measurable by its number of out-licensing contracts; therefore, every success means a figurehead and vice versa; every failure means accountability. In this context, the demo day is a visual example of how narratives serve a purpose and become a performance act for emotions around the artefact.

The idea's purpose is 'retold' and manufactured in this show, ultimately generating emotions that appeal to the audience. Making a good impression is essential because demo day becomes a sales stage to attract potential buyers, whether companies, insurance companies, or other financiers. For the teams of *Ellie* or *Feety*, the demo day involved prior practice with a theatre coach to tell a strategic and dramatic story that delivers a 'valuable' story in coherence with body language and emotions.

Ultimately, the emotions themselves become the product. Within the performance of a narrative to emotion, it is a process of accumulating many perspectives, from de-idealisation to functionalisation. In the process of gradual reduction, the emotions change their meaning. They are still world-making but have become flexible enough to serve a product or a market. *Emotions become functional companions and aim to reach or expand an audience, correspond to their ideas, or fulfil their wishes.*

8.2 ...Activity

Innovation is a collective term that arises from the self-image of our belief in progress. Through its broad application and the embedding of the modern belief in progress, it benefits from firm images and, through its imaginary form, permeates society, which, confronted with problems – whether individual or global – optimistically relies on the concept. It offers space for hopes, dreams, and wishes. The demand for the individual to make something innovative out of their imagination and ideas is tremendous. The idea cannot just remain an idea. The idea itself gains in value when it corresponds to a typical social idea of problem-solving. What we encountered are terms of increasing commodification that also permeate the imaginary and, thus, the emotionally inherent aspects. The imaginary and the creative have long since become an industry that has been taken up by corporate discourse, the DIY sector, and politics, whether in the call for more innovation and optimisation, an innovation union, or the idea of 'hacks'.

This realisation in connection with the reduction processes prompted a thought in me, especially in the last year of my work, that I have not been able to turn away from ever since: why is an idea initially built up and later reduced?

The multiplicity of an idea, the former vision, which certainly takes place individually at first, profits from the enrichment of many perspectives. Indeed, a shift in the focus of the problem would be more desirable, and yet plurality as such remains an enrichment. The one and the many ideas that become unified and perhaps ultimately indistinguishable in their iterative loops suffer from a certain point under capitalist market logic. The former enrichment of different ideals is reduced to a necessity, a profit. It is subjugated – to a system, a logic, a label.

Moreover, if I may now turn again to a theme of 'the one and the many', I would like to pose a question: why do we not enrich our ideas with what is already there but what we do not yet know? Why do we not enrich our perspective with that of others rather than reduce it again per our own logic? Why do we not start thinking of other logic and ontologies? This would not be innovative in the sense of the unprecedented, but it would nevertheless be new. It would be the turning away from the progress that can only justify itself through a perspective that merely knows what is inherent in it. The problems persist, and the next label is already known: *sustainability*. Often, we find both tied together, namely being innovative and sustainable. Yet, the perplexity persists in both as we always find the hierarchisation of perspectives, logic, ideas, desires, and evaluations. Perhaps with problems that exist now, both highly evident and pervasive, we can introduce solutions already there, even if they are not in 'our world'.

During this ethnography and writing process, I learned that problems could no more be observed in a sterile way than a solution can be a sterile answer. Through the described reduction, which is ultimately an expression of rationalisation, on the way to further development, the teams lost a great deal of what they initially recognised as valuable and what is rich at the beginning subsequently becomes increasingly sterile. This mode harbours the dangers of a 'sterile fantasy' (Illouz, 2017: 114) as reduction is the confirmation of removal from former intimacy. By this, I mean the discovery of a grievance, a problem. To recognise and engage with a problem requires closeness, empathy, and intimacy, and the discoverer invests effort and care when they want to solve the problem and get to its root. When other actors do this together with them, a network of intimacies and closeness is created; their exchange, as I have described, takes place in their moral economy, and through their solution, the problem is addressed. At this point, I would like to link the dangers of a sterile fantasy with what Wilkie et al. criticised about a future design constrained by risk aversion (Wilkie et al., 2017). This risk aversion or rigid view of problems, for the purpose of the more popular linearity and predictability, is precisely those reductions due to a known and represented logic that follows from the above. In this respect, proximity to society and to a problem and, at the same time, a fearlessness that is not oriented towards the rationalities of the present is what we need to innovate in a visionary and, above all, future-oriented way. Yet, these days, fears and worries are justified, but not because we do not know what is to come. Weather services, climate researchers, and military experts predict and calculate the world's future through all kinds of technological tools and around the clock. Instead, the drive for predictability has checkmated us to where we can only tremble. What prompts our feelings of insecurity? Is it the persistent uncertainty, which, contrary to expectations, is tinged with hope? Or is it our inertia signalling that we are downplaying legitimate concerns? Our emotions—concerns, joy, fears, and enthusiasm—remain the most reliable indicators of what requires attention and accomplishment. We need them in all their diversity to respond to our environment adequately; these moments of reaction offer space to create.

Nevertheless, I am rewriting the final lines of this concluding chapter as we can observe how emotions persistently become orchestrated and choreographed by several actors, e.g. the media, industrial companies and politicians that claim a certain emotional sovereignty of interpretation in everyday life. This idiosyncratic deprivation and recontextualisation of emotions not only signifies a misuse of authority but also the dehumanisation of human characteristics. Consequently, society is divested of its emotional autonomy and lacks interpretative sovereignty – a perilous circumstance for all. This alienation results in societal detachment from its core values and undermines opportunities for independent thinking, feeling and acting – all essential elements for autonomy, innovation and collaboration, and ultimately for creativity.

References

Adam, B., Beck, U., & van Loon, J. (2000). *The Risk Society and Beyond: Critical Issues for Social Theory*. Sage Publications.

Aghamanoukjan, A. (2012). Indikatoren des Neuen Innovation als Sozialmethodologie oder Sozialtechnologie? In I. Bormann, R. John, & J. Aderhold (Eds.), *Indikatoren des Neuen: Innovation als Sozialmethodologie oder Sozialtechnologie?* (pp. 227–250). Springer VS.

Akrich, M. (1992). The De-Scription of Technological Objects. In W. E. Bijker & J. Law (Eds.), *Shaping Technology/Building Society: Studies in Sociotechnical Change* (pp. 205–224). MIT Press.

Amabrush. (2019). *Amabrush - The 10 Second Toothbrush*. Kickstarter. Retrieved 23/11/21, 12:24 from https://www.kickstarter.com/projects/amabrush/amabrush-worlds-first-automatic-toothbrush/description

Amelang, K. (2012). Laborstudien. In S. Beck, J. Niewöhner, & E. Sørensen (Eds.), *Science and Technology Studies: eine sozialanthropologische Einführung*. transcript.

Archer, M. (2000). *Being Human: The Problem of Agency*. Cambridge University Press.

Assmann, J. (2011). *Cultural Memory and Early Civilization: Writing, Remembrance, and Political Imagination*. Cambridge University Press.

Bächtiger, A., Dryzek, J. S., Mansbridge, J. J., & Warren, M. (2018). *The Oxford Handbook of Deliberative Democracy*. Oxford University Press,. https://doi.org/http://dx.doi.org/10.1093/oxfordhb/9780198747369.001.0001

Ballon, P., Schuurman, D., & Blackman, C. (Eds.). (2015). *Living Labs: Concepts, Tools and Cases*. Emerald.

Barbalet, J. (2005). Smith's Sentiments (1759) and Wright's Passions (1601): The Beginnings of Sociology. *The British Journal of Sociology*, 56(2), 171–189. https://doi.org/https://doi.org/10.1111/j.1468-4446.2005.00054.x

Barbalet, J. (2006). Emotion. *Contexts*, 5(2), 51–53. https://doi.org/10.1525/ctx.2006.5.2.51

Bartel, C., & Garud, R. (2009). The Role of Narratives in Sustaining Organizational Innovation. *Organization Science*, 20, 107–117. https://doi.org/10.1287/orsc.1080.0372

Barthes, R. (1972). *Mythologies*. Selected and translated from the French by A. Lavers. Cape.
Bauer, R. (2017). Gescheiterte Innovationen als Gegenstand technikhistorischer Forschung. In W. Burr & M. Stephan (Eds.), *Technologie, Strategie und Organisation* (pp. 311–331). Springer VS. https://doi.org/10.1007/978-3-658-16042-5_16
Bausinger, H. (1958). Strukturen des alltäglichen Erzählens. 1(2), 239–254. https://doi.org/doi:10.1515/fabl.1958.1.2.239
Bausinger, H. (2016). *Alltägliches Erzählen*. Enzyklopädie des Märchens Online: Handwörterbuch zur historischen und vergleichenden Erzählforschung.
Beaney, M. (2005). *Imagination and Creativity*. Open University Worldwide.
Beck, U. (1986). *Risikogesellschaft: Auf dem Weg in eine andere Moderne* (1. Aufl., Erstausg. ed.). Suhrkamp.
Bernstein, R. J. (2010). *The Pragmatic Turn*. Polity.
Bijker, W. E., Hughes, T. P., & Pinch, T. (2012). *The Social Construction of Technological Systems: New Directions in the Sociology and History of Technology* (Anniversary ed.). MIT Press.
Bloch, E. (1980). *Abschied von der Utopie? Vorträge*. Suhrkamp.
Boden, M. A. (2004). *The Creative Mind: Myths and Mechanisms* (2nd ed.). Routledge.
Bogusz, T. (2009). Erfahrung, Praxis, Erkenntnis. Wissenssoziologische Anschlüsse zwischen Pragmatismus und Praxistheorie – ein Essay. *Sociologia Internationalis*, 2(47), 197–228. https://doi.org/10.3790/sint.47.2.197
Bolten, J. (2007). *Interkulturelle Kompetenz*. LZT.
Borghoff, B. (2018). Entrepreneurial Storytelling as Narrative Practice in Project and Organizational Development. In E. Innerhofer, H. Pechlaner, & E. Borin (Eds.), *Entrepreneurship in Culture and Creative Industries: Perspectives from Companies and Regions* (pp. 63–83). Springer International Publishing. https://doi.org/10.1007/978-3-319-65506-2_5
Bowman, D. M., Rip, A., & Stokes, E. (2017). *Embedding New Technologies into Society: A Regulatory, Ethical and Societal Perspective* (First edition. ed.). Pan Stanford Publishing.
Breidenstein, G., Hirschauer, S., Kalthoff, H., & Nieswand, B. (2020). *Ethnografie: die Praxis der Feldforschung* (3., überarbeitete Auflage ed.). UVK Verlag.
Brentano, F. (2015). *Psychology from an Empirical Standpoint*. Routledge.
Breuer, F., Muckel, P., & Dieris, B. (2019). *Reflexive Grounded Theory. Eine Einführung für die Forschungspraxis* (4. Auflage ed.) https://doi.org/10.1007/978-3-658-22219-2
Briken, K. (2006). Gesellschaftliche (Be-)Deutung von Innovation. In B. Blättel-Mink (Ed.), *Kompendium der Innovationsforschung* (1. Auflage ed., pp. 17–28). Springer VS. http://doi.org/10.1007/978-3-531-19971-9
Brooker, C. (2011). *Black Mirror* [Internet]. Netflix.

Bryan, L. A., & Tobin, K. (2019). *Critical Issues and Bold Visions for Science Education*. Brill | Sense. http://doi.org/10.1163/9789004389663

Bulkeley, H., Mai, L., Marvin, S., McCormick, K., Voytenko Palgan, Y., & Taylor and Francis. (2018). *Urban Living Labs: Experimenting with City Futures*. Routledge.

Bundesministerium für Wirtschaft und Energie, B. (2018). *Schlaglichter der Wirtschaftspolitik. Monatsbericht Januar 2019* https://www.bmwk.de/Redaktion/DE/Publikationen/Schlaglichter-der-Wirtschaftspolitik/schlaglichter-der-wirtschaftspolitik-01-2019.pdf?__blob=publicationFile&v=6

Bundesministerium für Wirtschaft und Energie, B. (2020). *From the Idea to Market Success* https://www.bmwk.de/Redaktion/EN/Publikationen/Technologie/from-the-idea-to-market-success.pdf?__blob=publicationFile&v=3

Byrne, R. M. J. (2005). *The Rational Imagination: How People Create Alternatives to Reality*. The MIT Press. https://doi.org/10.7551/mitpress/5756.001.0001

Chesky, B. (2018). How to Build Culture. In J. Løw (Ed.), *The Gurubook: Insights from 45 Pioneering Entrepreneurs and Leaders on Business Strategy and Innovation* (pp. 76–77). CRC Press, Taylor & Francis Group.

Ciaudo, J., Cismondi, F., Espahangizi, K., Gaupp, L., Hirschauer, S., Marten-Finnis, S., Niccolai, M., Pelillo-Hestermeyer, G., Reichardt, D., Sancho Höhne, M., & Ursula Oettl, B. (2021). *Diversity and Otherness: Transcultural Insights into Norms, Practices, Negotiations*. De Gruyter Open Poland.

Collins, H. M., & Evans, R. (2002). The Third Wave of Science Studies: Studies of Expertise and Experience. *Social Studies of Science, 32*(2), 235–296. https://doi.org/10.1177/0306312702032002003

Collins, H. M., & Pinch, T. (2014). *The Golem at Large: What You Should Know About Technology*. Cambridge University Press.

Comanducci, C., & Wilkinson, A. (2019). *Matters of Telling: The Impulse of the Story*. Leiden | Brill.

Comer, C. M., & Taggart, A. (2020). *Brain, Mind and the Narrative Imagination* (First ed.). Bloomsbury Publishing. https://doi.org/https:/doi.org/10.5040/9781350127838

Corsín Jiménez, A. (2011). Trust in Anthropology. *Anthropological Theory, 11*(2), 177–196. https://doi.org/10.1177/1463499611407392

Corsín Jiménez, A. (2014). Introduction. *Journal of Cultural Economy, 7*(4), 381–398. https://doi.org/10.1080/17530350.2013.858059

Cristea, I. A., Cahan, E. M., & Ioannidis, J. P. A. (2019). Stealth Research: Lack of Peer-Reviewed Evidence from Healthcare Unicorns. *Eur J Clin Invest, 49*(4), e13072– e13078. https://doi.org/10.1111/eci.13072

Curnow, R. C., & Moring, G. G. (1968). 'Project Sappho': A Study in Industrial Innovation. *Futures, 1*(2), 82–90. https://doi.org/10.1016/S0016-3287(68)80001-1

DAI-Labor. (2021). *DAI-Labor: About Us*. Retrieved 20/11/21; 15:18 from https://dai-labor.de/en/about-us/

Daston, L. (1995). The Moral Economy of Science. *Osiris, 10*, 3–24.
Daston, L., & Galison, P. (2007). *Objectivity*. Zone Books.
Davies, S. R. (2017). *Hackerspaces: Making the Maker Movement*. Polity Press.
Day, G. S., & Shea, G. P. (2018, 12/03/2022; 15:42). Grow Faster by Changing Your Innovation Narrative. *MIT Sloan Management Review, 60*(2). https://sloanreview.mit.edu/article/grow-faster-by-changing-your-innovation-narrative
de la Cadena, M., & Blaser, M. (2018). *A World of Many Worlds*. Duke University Press. https://doi.org/10.2307/j.ctv125jpzq
De Sousa, R. (1987). *The Rationality of Emotion*. MIT Press.
Deutschmann, C. (2020). Der Glaube der Finanzmärkte: Manifeste und latente Performativität in der Wirtschaft. In *Trügerische Verheißungen: Markterzählungen und ihre ungeplanten Folgen* (pp. 129–143). Springer VS. https://doi.org/10.1007/978-3-658-28582-1_8
Dewey, J. (1929). *The Quest for Certainty*. Minton, Balch & Company.
Dewey, J. (1934). *Art as Experience*. George Allen & Unwin.
Dickel, S. (2017). Irritierende Objekte. Wie Zukunft prototypisch erschlossen wird. *Behemoth, 10*(3), 171–190.
Dickel, S. (2019). *Prototyping Society – Zur vorauseilenden Technologisierung der Zukunft*. transcript.
Dixon, T. J., Connaughton, J. E., & Green, S. (2018). *Sustainable Futures in the Built Environment to 2050: A Foresight Approach to Construction and Development*. Wiley-Blackwell. https://doi.org/10.1002/9781119063834
Döveling, K., Konijn, E., Scheve, C. v., & ProQuest (Firm). (2010). *The Routledge Handbook of Emotions and Mass Media*. Routledge.
Durkheim, E. (1972). *Emile Durkheim: Selected Writings*. Cambridge University Press.
Durkheim, E. (2013). *The Division of Labour in Society* (S. Lukes, Trans.; Second ed.). Palgrave Macmillan.
Eco, U., Weaver, W., Bruegel, P., & Secondari, L. (1998). *Serendipities: Language and Lunacy*. Columbia University Press.
Elias, G. S., Garfield, R., Gutschera, K. R., Whitley, P., Zimmerman, E., & ProQuest. (2012). *Characteristics of Games*. MIT Press.
Escobar, A. (2020). *Pluriversal Politics: The Real and The Possible*. Duke University Press.
European Commission, E. (2010). *The 'Innovation Union' – Turning Ideas into Jobs, Green Growth and Social Progress* https://ec.europa.eu/commission/presscorner/detail/en/IP_10_1288
Fancher, R. E. (1977). Brentano's Psychology from an Empirical Standpoint and Freud's early metapsychology. *Journal of the History of the Behavioral Sciences, 13*, 207–227.
Färber, A., Ege, M., Binder, B., Audehm, K., & Althans, B. (2008). Kreativität. Eine Rückrufaktion.

Farias, I., & Wilkie, A. (2016). *Studio Studies: Operations, Topologies and Displacements*. Routledge.
Farson, R., & Keyes, R. (2003). *The Innovation Paradox: The Success of Failure, the Failure of Success*. Free Press.
Finance, Y. (2021, 31/08/2021). *Elizabeth Holmes: The 'Valley of Hype'. Behind the Rise and Fall of Theranos* https://www.youtube.com/watch?v=to2GSibbrv0
Fineman, S. (Ed.). (1993). *Emotion in Organizations*. Sage.
Fleck, L. (1935). *Entstehung und Entwicklung einer wissenschaftlichen Tatsache: Einführung in die Lehre vom Denkstil und Denkkollektiv. Mit mehreren Abbildungen*. Schwabe.
Fleck, L. (1980). *Entstehung und Entwicklung einer wissenschaftlichen Tatsache: Einführung in die Lehre vom Denkstil und Denkkollektiv*. Suhrkamp.
Flick, U. (2018). *The Sage Handbook of Qualitative Data Collection*. Sage Publications Ltd.
Florida, R. L. (2004). *The Rise of the Creative Class: And How it's Transforming Work, Leisure, Community and Everyday Life*. Basic Books.
Frantzeskaki, N., Hölscher, K., Bach, M., & Avelino, F. (2018). *Cocreating Sustainable Urban Futures: A Primer on Applying Transition Management in Cities*. Springer VS. https://doi.org/10.1007/978-3-319-69273-9
Frederiksen, M. (2016). Divided Uncertainty: a Phenomenology of Trust, Risk and Confidence. In S. Jagd & L. Fuglsang (Eds.), *Studying trust as process*. Edward Elgar Publishing. https://doi.org/10.4337/9781783476206.00008
Friedl, C. (2013). *Hollywood im journalistischen Alltag. Storytelling für erfolgreiche Geschichten. Ein Praxisbuch* http://doi.org/10.1007/978-3-658-00413-2
Gammerl, B. (2012). Emotional styles – Concepts and Challenges. *Rethinking History*, 16(2), 161–175. https://doi.org/10.1080/13642529.2012.681189
Gaut, B. (2003). Creativity and Imagination. In B. Gaut & P. Livingston (Eds.), *The Creation of Art: New Essays in Philosophical Aesthetics* (pp. 148–173). Cambridge University Press.
Godin, B. (2017). Why is Imitation not Innovation? In B. Godin & D. Vinck (Eds.), *Critical Studies of Innovation: Alternative Approaches to the Pro-Innovation Bias* (pp. 17–32). Edward Elgar Publishing.
Gordon, R. (1987). *The Structure of Emotions: Investigations in Cognitive Philosophy*. Cambridge University Press.
Gordon, S. L. (1981). The Sociology of Sentiments and Emotion. In M. Rosenberg & R. H. Turner (Eds.), *Social Psychology: Sociological Perspectives* (pp. 562–592). Basic Books.
Graeber, D. (2018). *Bullshit Jobs: A Theory*. Penguin Books.
Guggenheim, M. (2010). The Long History of Prototypes. *Limn*, 1(1).
Guggenheim, M. (2012). Laboratizing and De-Laboratizing the World: Changing Sociological Concepts for Places of Knowledge Production. *History of the Human Sciences*, 25(1), 99–118. https://doi.org/10.1177/0952695111422978

Guggenheim, M. (2014). From Prototyping to Allotyping. *Journal of Cultural Economy*, 7(4), 411–433. https://doi.org/10.1080/17530350.2013.858060

Harris, M., & Rapport, N. (2015). *Reflections on Imagination: Human Capacity and Ethnographic Method*. Ashgate.

Higgins, T. C. (1975). *Success and Failure in Innovation and the Implications for R & D Management and Choice*. University College Dublin.

Hilgartner, S., Miller, C., & Hagendijk, R. (2015). *Science and Democracy: Making Knowledge and Making Power in the Biosciences and Beyond*. Routledge, Taylor & Francis Group.

Hinkel, J., Mangalagiu, D., Bisaro, A., & Tàbara, J. D. (2020). Transformative Narratives for Climate Action. *Climatic Change*, 160(4), 495–506. https://doi.org/10.1007/s10584-020-02761-y

Hintikka, M. B., & Hintikka, J. (1986). *Investigating Wittgenstein*. Blackwell.

Hitzer, B., & Gammerl, B. (2013). Wohin mit den Gefühlen? Vergangenheit und Zukunft des Emotional Turn in den Geschichtswissenschaften. *Auf der Jagd nach Gefühlen*, 24. Jg.(3), 31–40.

Hochschild, A. R. (2012). *The Managed Heart: Commercialization of Human Feeling* (Updated with a new preface. ed.). University of California Press.

Hoffman, U., & Marz, L. (1996). *Visions of Technology: Social and Institutional Factors Shaping the Development of New Technologies*. St. Martin's Press.

Hofstede, G. H., & Hofstede, G. (2001). *Culture's Consequences: Comparing Values, Behaviors, Institutions and Organizations Across Nations*. SAGE Publications.

Humboldt, A. (1845). *Kosmos : Entwurf einer physischen Weltbeschreibung* (1. ed., Vol. I). Cotta.

Husserl, E., & Ströker, E. (2012). *Die Krisis der europäischen Wissenschaften und die transzendentale Phänomenologie: eine Einleitung in die phänomenologische Philosophie*. Meiner.

Hutter, M. K., Hubert; Rammert, Werner; Windeler, Arnold. (2016). Innovationsgesellschaft heute. Die reflexive Herstellung des Neuen. In W. Rammert, A. Windeler, H. Knoblauch, & M. Hutter (Eds.), *Innovationsgesellschaft heute Perspektiven, Felder und Fälle* (pp. 15–38). Springer VS. http://doi.org/10.1007/978-3-658-10874-8

Huxley, A. (1932). *Brave new World : a novel* (3. impr. ed.). Chatto & Windus.

Illouz, E. (2007). *Cold Intimacies: The Making of Emotional Capitalism*. Wiley.

Illouz, E. (2017). *Emotions as Commodities: Capitalism, Consumption and Authenticity*. Taylor & Francis.

Ingold, T. (2000). *The Perception of the Environment: Essays on Livelihood, Dwelling and Skill*. Routledge.

Ioannidis, J. P. (2015). Stealth Research: Is Biomedical Innovation Happening Outside the Peer-Reviewed Literature? *JAMA*, 313(7), 663–664. https://doi.org/10.1001/jama.2014.17662

Irwin, A. (1995). *Citizen Science: A Study of People, Expertise and Sustainable Development*. Routledge.
Ishiguro, K. (2021). *Klara and the Sun* (Main ed.). Faber & Faber.
James, W. (1902). *The Principles of Psychology* (Vol. 1). Macmillan.
James, W. (1922). *Pragmatism : A New Name for Some Old Ways of Thinking. Popular Lectures on Philosophy by William James*. Longmans, Green and Co.
James, W. (2006). *Pragmatismus und radikaler Empirismus* (C. Langbehn, Ed. 1. Aufl. ed.). Suhrkamp.
Jasanoff, S., & Kim, S.-H. (2015). *Dreamscapes of Modernity: Sociotechnical Imaginaries and the Fabrication of Power*. The University of Chicago Press.
Jelinek, M., & Schoonhoven, C. B. (1990). *The Innovation Marathon: Lessons from High Technology Firms*. Jossey-Bass Publishers.
John, R. (2012). Erfolg als Eigenwert der Innovation. In I. Bormann, R. John, & J. Aderhold (Eds.), *Innovation und Gesellschaft, Indikatoren des Neuen. Innovation als Sozialmethodologie oder Sozialtechnologie?* (pp. 77–96). Springer VS. http://dx.doi.org/10.1007/978-3-531-94043-4
Johnson, S. (2010). *Where Good Ideas Come from: The Natural History of Innovation*. Riverhead Books.
Kelty, C., Jiménez, A. C., Marcus, G. E., & al., e. (2010). *Prototyping Prototyping Prototyping Prototyping*. Anthropological Research on the Contemporary ARC Studio, ARC Studio Madrid.
Keyson, D. V., Guerra-Santin, O., & Lockton, D. (2016). *Living Labs: Design and Assessment of Sustainable Living*. Springer International Publishing.
King, N. (2000). *Memory, Narrative, Identity: Remembering the Self*. Edinburgh University Press.
Kingdon, M. (2013). *The Science of Serendipity: How to Unlock the Promise of Innovation in Large Organisations*. Wiley, a John Wiley & Sons Ltd, Publication.
Knorr-Cetina, K. (1997). Sociality with Objects: Social Relations in Postsocial Knowledge Societies. *Theory, Culture & Society, 14*(4), 1–30. https://doi.org/10.1177/026327697014004001
Krüger, A. K., & Reinhart, M. (2016). Wert, Werte und (Be)Wertungen. Eine erste begriffs- und prozesstheoretische Sondierung der aktuellen Soziologie der Bewertung. *Berliner Journal für Soziologie, 26*(3), 485–500. https://doi.org/10.1007/s11609-017-0330-x
Krüger, A. K., & Reinhart, M. (2017). Theories of Valuation – Building Blocks for Conceptualizing Valuation between Practice and Structure. *Historical Social Research, 42*, 263–285.
Krüger, A. K., & Reinhart, M. (2018). Emotional Value Attribution and Comparative Value Assessment–Analytical Elements for a Sociology of Valuation and Evaluation. https://doi.org/10.31235/osf.io/huwk3

Kuenkel, P. (2018). *Stewarding Sustainability Transformations: An Emerging Theory and Practice of SDG Implementation*. Springer International Publishing.

Latour, B. (1987). *Science in Action: How to Follow Scientists and Engineers Through Society*. Open University Press.

Latour, B. (1993). *We Have Never Been Modern* (C. Porter, Trans.). Harvard University Press.

Latour, B. (1994). On Technical Mediation. *Common Knowledge, 3*(2), 29–64.

Latour, B. (1996). *Aramis or the Love of Technology*. Harvard University Press.

Latour, B. (2004). Why Has Critique Run out of Steam? From Matters of Fact to Matters of Concern. *Critical Inquiry, 30*(2). https://doi.org/https://doi.org/10.1086/421123

Latour, B. (2010). *On the Modern Cult of the Factish Gods*. Duke University Press.

Law, J. (2004). *After method: Mess in Social Science Research* (1. publ. ed.). Routledge.

Law, J., & Mol, A. (1995). Notes on Materiality and Sociality. *The Sociological Review, 43*(2), 274–294. https://doi.org/10.1111/j.1467-954X.1995.tb00604.x

Lazarus, R. S. (1991). Cognition and motivation in emotion. *The American psychologist, 46 4*, 352–367.

Levi-Strauss, C. (2021). *Wild Thought: A New Translation of 'La Pensée Sauvage'* (J. Mehlman & J. Leavitt, Trans.). University of Chicago Press.

Løw, J. (2018). *The Gurubook: Insights from 45 Pioneering Entrepreneurs and Leaders on Business Strategy and Innovation*. CRC Press, Taylor & Francis Group.

Luhmann, N., Baecker, D., & Gilgen, P. (2013). *Introduction to Systems Theory*. Polity.

Maasen, S. (2019). Digitale Technologien, ihr Unbewusstes, ihre Gesellschaft: Psychoanalyse als Gegenwissenschaft? In E. Frick, A. Hamburger, & S. Maasen (Eds.), *Psychoanalyse in technischer Gesellschaft Streitbare Thesen* (1. Auflage ed., pp. 191–202). Vandenhoeck & Ruprecht GmbH & Co. KG, V&R unipress GmbH. https://doi.org/https://doi.org/10.13109/9783666403873

Mackintosh, S. P. M. (2021). *Climate Crisis Economics* (1st ed.). Routledge.

Mannik, L., & McGarry, K. (2017). *Practicing Ethnography: A Student Guide to Method and Methodology*. University of Toronto Press.

Marres, N. (2012). *Material Participation: Technology, the Environment and Everyday Publics*. Palgrave Macmillan.

Marx, K. (1887). *Capital: A Critical Analysis of Capitalist Production*. De Gruyter.

Meadows, D., Meadows, D., Randers, J., & Behrens III, W. W. (1972). *The Limits to Growth: A Report for the Club Of Rome's Project on the Predicament of Mankind*. Universe Books.

Meissner, D., Polt, W., & Vonortas, N. S. (2017). Towards a Broad Understanding of Innovation and its Importance for Innovation Policy. *The Journal of Technology Transfer, 42*(5), 1184–1211. https://doi.org/10.1007/s10961-016-9485-4

Merton, R. K. (1968). The Matthew Effect in Science. *Science, 159*, 56 – 63.

Mielke, F. (2021). *Steps towards a Mindful Organisation: Developing Mindfulness to Manage Unexpected Events*. Springer Fachmedien Wiesbaden.

Miner, E. N. M. (1987). *Robo Cop* A. Schmidt.

Mol, A. (2008). *The Logic of Care: Health and the Problem of Patient Choice*. Routledge.

Morse, J. M., Bowers, B. J., Charmaz, K., Clarke, A. E., Corbin, J., Porr, C. J., & Stern, P. N. (2021). *Developing Grounded Theory: the Second Generation* (Second ed.). Routledge.

Moultrie, J., Nilsson, M., Dissel, M., Haner, U.-E., Janssen, S., & Van der Lugt, R. (2007). Innovation Spaces: Towards a Framework for Understanding the Role of the Physical Environment in Innovation. *Creativity and Innovation Management*, 16(1), 53–65. https://doi.org/https://doi.org/10.1111/j.1467-8691.2007.00419.x

Müller, M. (2013). How Innovations Become Successful through Stories. In A. P. Müller & L. Becker (Eds.), *Narrative and Innovation: New Ideas for Business Administration, Strategic Management and Entrepreneurship* (pp. 139–149). Springer VS. https://doi.org/10.1007/978-3-658-01375-2_9

Münkler, H. (2009). *Die Deutschen und ihre Mythen* (1. Aufl. ed.). Rowohlt.

Museum, C. H. (2014). *CHM Revolutionaries: Theranos Founder & CEO Elizabeth Holmes in Conversation with Michael Krasny*. Computer History Museum. https://www.youtube.com/watch?v=uJDc4tOU3zo&list=PLvJfo4my4JJKjIUPoMdV_zy8Bum9_1RbB&ab_channel=ComputerHistoryMuseum

N.A. (2021a). *Bosch IoT Campus: Über uns*. Bosch IoT Campus. Retrieved 20/11/21, 15:32 from https://bosch.io/de/ueber-uns/standorte/berlin/

N.A. (2021b). *Happy Lab Berlin*. Retrieved 20/12/2021, 15:28 from https://www.happylab.de/de_ber/

N.A. (2022). *How to Build a Compelling Brand Story*. Momentive Europe UC. Retrieved 05/12/2022 from https://www.surveymonkey.de/market-research/resources/how-to-write-great-brand-story/

Nold, C. (2018). Turning Controversies into Questions of Design: Prototyping Alternative metrics for Heathrow Airport. In. Mattering Press.

Nowotny, H. (2010). Vernunft und Innovation. In *Innovationskultur – zur Produktion neuen Wissens* (pp. 415–426). Brill | Fink. https://doi.org/https://doi.org/10.3096 5/9783846750735_046

Nussbaum, M. C. (2003). *Upheavals of Thought: The Intelligence of Emotions*. Cambridge University Press.

Oatley, K. (1993). *Social construction in emotions* The Guilford Press.

Parker, S. C. (2011). Intrapreneurship or Entrepreneurship? *Journal of Business Venturing*, 26(1), 19–34. https://doi.org/https://doi.org/10.1016/j.jbusvent.2009.07.003

Pfotenhauer, S. M. U. (2017). Innovation & Society: The Diversity of Innovation Practice. *EASST Review*, 36(1), 33–36.

Picard, R. (2017). *Co-Design in LIving Labs for Healthcare and Independent Living: Concepts, Methods and Tools*. ISTE Ltd. Wiley. https://doi.org/10.1002/9781119388746

Puig de la Bellacasa, M. (2017). *Matters of Care: Speculative Ethics in More than Human Worlds*. University of Minnesota Press.

Qmarkets. (2021). *Discover Breakthrough Ideas with an Employee Innovation Program*. Retrieved 20/11/2021, 18:41 from https://www.qmarkets.net/use-cases/by-business-challenge/employee-innovation-program/

Reckwitz, A. (2017). *The Invention of Creativity: Modern Society and the Culture of the New* (S. Black, Trans.). Polity.

Reinhart, M. (2012). Wissenschaft und Wirtschaft: Von Entdeckung zu Innovation. In S. Maasen, M. Kaiser, M. Reinhart, & B. Sutter (Eds.), *Handbuch Wissenschaftssoziologie* (pp. 365–378). Springer Fachmedien Wiesbaden. https://doi.org/10.1007/978-3-531-18918-5_29

Reinhart, M. (2016). Rätsel und Paranoia als Methode. Vorschläge zu einer Innovationsforschung der Sozialwissenschaften. In A. Froese, D. Simon, & J. Böttcher (Eds.), *Sozialwissenschaften und Gesellschaft. Neue Verortungen von Wissenstransfer* (pp. 342). transcript Verlag.

Reinhart, M., Krüger, A. K., & Hesselmann, F. (2019). Nach der Bewertung ist vor der Bewertung – Sichtbarkeit und Emotionalität als verbindende Elemente von Bewertungsprozessen. In S. Nicolae, M. Endreß, O. Berli, & D. Bischur (Eds.), *(Be)Werten. Beiträge zur sozialen Konstruktion von Wertigkeit* (pp. 125–145). Springer VS. https://doi.org/10.1007/978-3-658-21763-1_6

Rheinberger, H.-J. (2014). Über Serendipität – Forschen und Finden. In O. Budelacci, G. Boehm, G. Wildgruber, & E. Alloa (Eds.), *Imagination* (pp. 233–243). Brill | Fink. https://doi.org/https://doi.org/10.30965/9783846756232_012

Ricœur, P. (1988). *Zeit und Erzählung*. Fink.

Ricœur, P. (1995). *Figuring the Sacred: Religion, Narrative, and Imagination* (M. I. Wallace, Trans.). Fortress Press.

Rip, A. E., Misa, T. J. E., & Schot, J. E. (1995). *Managing Technology in Society: The Approach of Constructive Technology Assessment: International Workshop: Papers*. Pinter Publishers.

Robinson, J. (2005). *Deeper than Reason: Emotion and its Role in Literature, Music, and Art*. Oxford University Press.

Roddenberry, G. (1966). *Star Trek* [TV Series]. D. Productions; NBC.

Roese, N. J., & Olson, J. M. (1995). Functions of Counterfactual Thinking. In N. J. Roese & J. M. Olson (Eds.), *What Might Have Been: The Social Psychology of Counterfactual Thinking* (pp. 169–197). Erlbaum.

Russell, B. (1975). *Philosophie des Abendlandes: ihr Zusammenhang mit der politischen und der sozialen Entwicklung*. Europaverlag.

Sartre, J. P. (2015). *Sketch for a Theory of the Emotions*. Taylor & Francis.

Savransky, M. (2021a). *Around the day in eighty worlds : politics of the pluriverse*. Duke University Press.

Savransky, M. (2021b). The Pluralistic Problematic: William James and the Pragmatics of the Pluriverse. *Theory, Culture & Society*, 38(2), 141–159. https://doi.org/10.1177/0263276419848030

Schneider, I. (2016). Über die emotionale Kompetenz der Europäischen Ethnologie/ Empirischen Kulturwissenschaft/ Kulturantrhopologie. Zur Einführung. In M. S. Beitl, Ingo (Ed.), *Buchreise Österreichischen Zeitschrift für Volkskunde* (Vol. Neue Serie, Band 27).

Schumpeter, J. A. (1942). *Capitalism, Socialism, and Democracy*. Harper & Brothers.

Schumpeter, J. A., Röpke, J., & Stiller, O. (2006). *Theorie der wirtschaftlichen Entwicklung* (Nachdruck der 1. Auflage von 1912 ed.). Duncker & Humblot.

Seidenschnur, T. (2019). The Logic of Innovation. A Study on the Narrative Construction of Intrapreneurial Groups in the Light of Competing Institutional Logics. *Historical Social Research / Historische Sozialforschung*, 44(4 (170)), 222–249.

Shelley, M. (1897). *Frankenstein or The modern Prometheus*. Gibbings.

Shils, E. (1981). *Tradition*. University of Chicago Press.

Simmel, G. (1905). A Contribution to the Sociology of Religion. *American Journal of Sociology*, 11(3), 359–376. https://doi.org/10.1086/211407

Smircich, L. (1983). Concepts of Culture and Organizational Analysis. *Administrative Science Quarterly*, 28(3), 339–358. https://doi.org/10.2307/2392246

Smorti, A. (2020). *Telling to Understand: The Impact of Narrative on Autobiographical Memory*. Springer VS.

Solomon, R. C. (1993). *The Passions: Emotions and the Meaning of Life*. Hackett.

Solomon, R. C. (2004). Emotions, Thoughts, and Feelings: Emotions as Engagements with the World. In R. C. Solomon (Ed.), *Thinking About Feeling: Contemporary Philosophers on Emotions* (pp. 1–18). Oxford University Press.

Spielberg, S. (2022). *Minority Report* [Cinema]. B. Curtin, J. d. Bont, G. R. Molen, & W. F. Parkes;

Star, S. L. (1992). The trojan door: Organizations, work, and the "open black box". *Systems practice*, 5(4), 395–410. https://doi.org/10.1007/BF01059831

Star, S. L., & Griesemer, J. R. (1989). Institutional Ecology, 'Translations' and Boundary Objects: Amateurs and Professionals in Berkeley's Museum of Vertebrate Zoology, 1907–39. *Social Studies of Science*, 19(3), 387–420. https://doi.org/10.1177/030631289019003001

Strathern, M. (2005). Robust Knowledge and Fragile Futures. In A. O. S. J. Collier (Ed.), *Global Assemblages: Technology, Politics, and Ethics as Anthropological Problems* (pp. 464–481). Blackwell Publishing. https://doi.org/10.1002/9780470696569.ch24

Suchman, M. C. (1995). Managing Legitimacy: Strategic and Institutional Approaches. *Academy of Management Review*, 20(3), 571–610. https://doi.org/10.5465/amr.1995.9508080331

Thygesen, T. (2018). Why Culture is Just as Important for Tech Startups as Tech. In J. Løw (Ed.), *The Gurubook: Insights from 45 Pioneering Entrepreneurs and Leaders on Business Strategy and Innovation* (pp. 78–79). CRC Press, Taylor & Francis Group, https://www.taylorfrancis.com/books/9781315123752

Tronto, J. C. (1993). *Moral Boundaries: A Political Argument for an Ethic of Care*. Routledge.

Urry, J. (2016). *What is the Future?* Polity.

Vallery-Radot, R. (1926). *The Life of Pasteur*. Garden City Publishing Company.

Villanueva, J. (2012). Does it Matter How You Tell it? On How Entrepreneurial Storytelling Affects the Opportunity Evaluations of Early-Stage Investors. *Frontiers of entrepreneurship research*, 33, 1.

Wilkie, A., Savransky, M., & Rosengarten, M. (2017). *Speculative research: The Lure of Possible Futures*. Routledge.

Williams, A. (2022). *The Power of the Founder's Story. And How to Write Yours, By Us*. Cohesive. Retrieved 16/12/2022; 11:02 from https://wearecohesive.com/stories/articles/founders-story-and-how-to-write-yours/

Wills, I. (2019). Innovation Must Fail. In *Thomas Edison: Success and Innovation through Failure* (pp. 81–96). Springer VS. https://doi.org/10.1007/978-3-030-29940-8_5

Wissenschaftlicher Dienst des Deutschen Bundestages, W. (2018). *Reallabore, Living Labs und Citizen Science-Projekte in Europa*. Berlin: Wissenschaftliche Dienste des Deutschen Bundestages Retrieved from https://www.bundestag.de/resource/blob/563290/9d6da7676c82fe6777e6df85c7a7d573/wd-8-020-18-pdf-data.pdf

Wittgenstein, L. (1977). *Philosophische Untersuchungen*. Suhrkamp.

Wynne, B. (1998). May the Sheep Safely Graze? A Reflexive View of the Expert-Lay Knowledge Divide. In S. Lash, B. Szerszynski, & B. Wynne (Eds.), *Risk, Environment and Modernity: Towards a New Ecology* (pp. 44–83). Sage Publications. https://doi.org/10.4135/9781446221983

Yaqub, O. (2018). Serendipity: Towards a Taxonomy and a Theory. *Research Policy*, 47(1), 169–179. https://doi.org/https://doi.org/10.1016/j.respol.2017.10.007